Internet Research for Middle School Students

ENVIRONMENTAL SCIENCE ON THE NET

THE GLOBAL THINKING PROJECT

By

JACK HASSARD AND JULIE WEISBERG

Georgia State University

Agnes Scott College

Good Year Books

An Imprint of Addison-Wesley Educational Publishers, Inc.

Dedication

This book is dedicated to Dr. Jenny Springer, Associate Superintendent of Instruction of DeKalb County Schools (and former principal of Dunwoody High School), Decatur, Georgia, U.S.A., and to the memory of Mr. Vadim Zhudov, Director of Experimental Gymnasium School 710, Moscow, Russia. These dedicated educators provided the emotional support and intellectual guidance to foster an environment in which educators and students worked together to develop the Global Thinking Project, upon which this book *Environmental Science on the Net: The Global Thinking Project* is based. Zhudov and Springer supported the first student and educator exchanges between the United States and Russia (then the Soviet Union), and opened their schools to the experimental work required by the Global Thinking Project. From 1988 to the present, these educators supported more than six student exchanges and ten educator exchanges.

Good Year Books

are available for most basic curriculum subjects plus many enrichment areas. For more Good Year Books, contact your local bookseller or educational dealer. For a complete catalog with information about other Good Year Books, please write:

Good Year Books
299 Jefferson Road
Parsippany, NJ 07054-0480

Book Design: Meyers Design

Cover: Meyers Design

Developmental Editors: Susan Buntrock and Laura Strom

Production/Manufacturing Coordinator: Roxanne Knoll

Printed in the United States of America.

ISBN 0-673-57732-5

1 2 3 4 5 6 7 8 9 ML 06 05 04 03 02 01 00 99

Editor's note: Names of service providers and equipment providers are given for reference purposes only and do not constitute an endorsement of any particular service equipment by the authors or the publisher or its agents.

Table of Contents

Introduction

This book, *Environmental Science on the Net: The Global Thinking Project*, has been written to provide educators with solid information on how to provide their students with opportunities for global communication as well as environmental study. This material was designed to be used in conjunction with the Internet. You might think of the Global Thinking Project (GTP) curriculum as a "Web-assisted teaching program." In other words, you will use a variety of tools available through the Internet to foster communication among other educators and students, in conjunction with local environmental investigations. You will use e-mail to send messages to colleagues, and your students will use it to establish communication with their peers around the world. You might want to use video conferencing (which is very inexpensive and easy to use) to carry out Internet conferences. You will use a browser not only to search the Internet for relevant information about the environment, but to communicate, to share data, to look at other students' data, and to see images of students and teachers using the GTP.

A "global thinking educator" combines human values of concern and appreciation of others with a strong interest in science and social issues that affect humankind. In the GTP, science serves people. It is used to help solve real problems facing communities around the world. We welcome you to the GTP and know that you and your students will be able to construct personally meaningful experiences as you use our materials.

Brief History of the Global Thinking Project

The GTP is a grass-roots environmental education project conceived in seminar rooms and classrooms in Moscow and St. Petersburg, Russia and Atlanta, Georgia.[1] The project owes its existence to the efforts of a small group of American and Russian classroom teachers who were interested in working together to learn about each other, how they teach, and how to improve the quality of learning in their classrooms. Building on a network of relationships established during the 1980s, primarily through the efforts of the Association for Humanistic Psychology, these teachers visited and taught in each others' classrooms and participated in teaching and learning seminars in each others' countries. In the fall of 1989, they met in St. Petersburg and agreed to work together to develop teaching materials that would:

[1] Hassard, J. "Teaching students to think globally." *Journal of Humanistic Psychology*, 37(1), 24–63, 1997.

1. Empower students and teachers to get involved with important global problems and concerns.
2. Introduce students to collaborative methods and strategies of inquiry that could be used to solve problems locally, and provide the knowledge and technological means needed to deal with problems globally.
3. Develop computer literacy in students that would allow them to use microcomputers as a telecommunications tool to collaborate with counterparts in other nations.[2]

The first version of these teaching materials was developed in 1989–1990, based on curriculum themes and strands identified by this original group of teachers. Since then, other educators from Russia and the United States, as well as from Spain, Australia, the Czech Republic, and Singapore have made important contributions. The activities included in these materials have been field-tested across many cultures in Grades 5–12.

How This Book Is Organized

The teaching materials contained in this guide provide a framework for teachers in different cultures to engage their students in local environmental research projects, as well as to collaborate with their counterparts in other cities, states, and countries through the GTP's home page (www.gtp.org). The book is divided into ten chapters:

Chapter 1

"Using the Internet to Collaborate Globally" shows you how to incorporate the resources of the World Wide Web into projects.

Chapter 2

"Project Hello" introduces students to cross-cultural communication and provides an opportunity for them to practice using the tools provided in the Global Thinking home page for research and data collection and exchange.

Chapters 3–8

These chapters include frameworks for environmental inquiry projects that may be completed in any order. Each framework includes suggestions for incorporating Internet resources into the inquiry.

Chapter 9

These Extension Activities, contributed by participating teachers, provide additional activities related to the Global Thinking Project.

Chapter 10

The Student Log consists of reproducible pages for recording data and observations as students conduct the environmental inquiry projects.

[2] Ibid. p. 26.

Using the GTP Materials

The materials in this book can be used in a number of different ways. They are meant to be a framework for planning environmental investigations with your students. Each project is accompanied by a planning guide, which includes advance planning alerts, suggested times for each activity, and sources for hands-on materials that you will need to carry out the monitoring activities. Please read these carefully before beginning a project, since some activities need a week or two of lead time for preparation of materials.

In Conjunction with the GTP Web Site

The GTP materials were designed to be used with Internet connections among teachers and students in different schools. In addition to the printed materials you will find here, you will also be directed to the GTP Web site (www.gtp.org).

If you plan to use the printed materials in connection with the Internet, you should use your browser to visit the GTP Web site. Here you will be prompted to join the GTP. Once you join, you will be entered into the GTP database, and other teachers and schools will identify you as a GTP school and teacher. Take some time to browse around the GTP Web site. The most important resources in the site will probably be the home pages for each of the projects that are developed in this book. The home pages will provide you with a virtual place to work, to send information, to retrieve information from other schools, and to post discussion comments. They'll also provide important links related to the content of the project.

If you use the program in this way, you will not only be surfing the Net for information about environmental science and issues, but sharing the data you collect in your environment with other schools by submitting it to the GTP database. You will also be retrieving data and participating in Web discussions. You might also want to try to take part in video conferences using the software suggested in Chapter 1. In these ways, you will be using the Web to make your classroom a hub, connecting your students to the world, and as an assistant to help you make your teaching interesting, timely, and relevant.

You can also use our materials in such a way that the Web is used in a more limited fashion. For example, you might want to use the home pages for the resources that we have identified, but not want to be totally involved in data sharing and retrieval. Or, it might be that you want to focus on local problems and use the tools and activities that we have developed to involve your students in local environmental issues.

Another example of limited use of the Web would be if you decide to try to find one other teacher or a group of students in another region of your country or in the world who want to collaborate on one of the GTP projects. In this case, you can use our Web home pages and e-mail to carry out an interesting global project.

As Stand-Alone Projects

You might be in a situation where you have very limited access to the Web, but you are interested in involving your students in environmental activities. Using the hands-on science materials suggested for each of the projects, you will be able to carry out an interesting environmental research program. When you get Web access, you will be ready to join in because you'll already be familiar with the content of the GTP.

The Global Thinking Student Log

The Global Thinking Student Log contains reproducible pages so students may keep observations, data, and other types of records for each project. The data tables on these pages are formatted to correspond with the data-sharing tables available on the GTP Web site. We suggest that you make a journal for each team of students in which to keep a record of its work during a project. Student journals will be one important source of assessment data for you.

Where to Begin

You should begin by visiting the GTP Web site at www.gtp.org. Take some time to explore this site, and if you decide you want to join the project and participate in its various activities and events, simply fill out the form on the "Join" link on the GTP home page.

GTP Web Site Home Page

To begin your work on the GTP, visit the site and explore the resources.

You should begin your work on the GTP with Project Hello. This will introduce you to the project and help your students understand the significance of working on a global science project. From that point on you can select the projects you wish to do from Chapters 3–8.

The GTP Planning Calendar

Although you can use the materials in the GTP in any sequence you wish, teachers typically followed the schedule outlined below during the period of development of these materials. During a year, most teachers do three or four projects, selecting one or more projects from Phases I and II, and joining in on the activities for Project Earth Month.

Phase I: September–December	Phase II: January–March	Phase III: April–May
Project Hello Project Clean Classroom Project Clean Air	Project Solid Waste Project Water Watch Project Soil	Project Earth Month

The Global Thinking Curriculum

What Is Global Thinking?

Ask your students what "global thinking" means to them. We did and here are some responses:

"Global thinking is thinking how our actions and reactions affect all the world." (female, age 13)

"Thinking about the world as a whole, as one, not separated by continental boundaries or individual politics." (male, age 13)

"I think thinking globally is when the Earth's people think as one about the Earth's problems and take responsibility for their share in the creation of those problems." (male, age 12)

"Thinking globally means thinking with and about the whole world." (male, age 13)

"Global thinking is when everyone on the Earth thinks of each other as one." (female, age 13)

"Thinking of the world as a whole with differences, but not divided." (male, age 15)

Global thinking is a new way of thinking. It represents a shift in thinking from a mechanistic, hierarchical model of science to a new, more flexible model based on inquiry and interdependence. It includes both the ability to anticipate the consequences of current and future actions and to participate in developing solutions to local and global environmental problems.

Global Thinking Contrasted with a Traditional Model of Thinking

The Traditional Model	The Global Thinking Model
• Traditional, mechanized thinking	• Innovative, flexible thinking
• Individualistic—although students may at times work together in groups, interdependence typically is not a goal.	• Cooperative—students work collaboratively in small teams to think and take action together.
• Dependence—teacher-directed instructional model establishes a dependent social system.	• Interdependence—a synergic system is established in groups within a classroom, and within global communities of practice.
• Hierarchical—choice made for you; rarely do students choose content or methodology for their investigations.	• Right to choose—students are involved in making choices including problem and topic selection, as well as solutions; reflects the action processes of grassroots organizations.
• Literacy—the emphasis is on knowing facts, skills, and concepts.	• Literacy—this is a new literacy insofar as "knowledge" relates to human needs, the needs of the environment, and the social needs of the Earth's population and other living species.
• Content—the emphasis is on acquiring the right body of knowledge.	• Content—the emphasis is on anticipation and participation, inquiry, learning how to learn, and how to ask questions.
• Learning encourages recall, and is analytical and linear.	• Learning encourages creative thinking, and is holistic and intuitive.

The Content of Global Thinking Project

The Global Thinking Project integrates environmental education, cooperative learning, and technology, enabling your students to become part of a community of global thinkers. They will look at the environment from local as well as global perspectives, learn how to investigate environmental problems, and be exposed to ways to take action on the problems they consider to be important.

From the student's perspective, the concept of global thinking is explored through a series of projects. Each project is a module of learning focusing on one aspect of the environment. In these projects, students are active learners, constructing their own knowledge about the environment. We have not only identified key environmental questions that students could explore, but also learning processes (such as collecting data, communicating, and inferring), and themes (such as energy, systems, interactions) that students might utilize to construct knowledge. The scope and sequence of the GTP below provides a content analysis of the curriculum.

Global Thinking Project Scope and Sequence Chart

Project	Learning Processes	Themes	Objectives	Environmental Education Focus of the Project
Project Hello	Collecting Data Communicating Comparing	Energy Systems and Interactions Scale Interdependence	Find out what is global thinking. Learn how to use the Internet to send e-mail. Evaluate global environmental problems.	What is global thinking and how is it related to environmental education? Is Earth a living system? What are the most pressing environmental problems facing humankind?
Project Green Classroom	Collecting data Communicating Comparing	Energy Systems and Interactions Scale Interdependence	Describe the physical, biological, and geographical characteristics of their environment.	Evaluate the environmental quality of your classroom. What is our environment like? How does it compare to the environment of other schools? How can we improve the environment of our school and classroom?
Project Clean Air	Collecting data Communicating Inferring Applying	Systems and Interactions Patterns of Change Interdependence Energy	Investigate patterns of atmospheric pollution. Learn techniques to monitor particulates and lower atmospheric ozone. Design a research project based on questions and inquiries about air quality.	What do you know about atmospheric pollution? Why is atmospheric monitoring important to our understanding of weather and climate? How does atmospheric pollution (particulate matter and ozone) in our community compare with other communities?

Global Thinking Project Scope and Sequence Chart

Project	Learning Processes	Themes	Objectives	Environmental Education Focus of the Project
Project Solid Waste	Collecting data Communicating Categorizing Inferring Applying	Patterns of Change Systems and Interactions Energy Stability Interdependence	Classify and measure solid waste. Identify ways to reduce solid waste. Determine rates of decomposition of solid waste. Evaluate solid-waste disposal methods.	Why is it important to study solid-waste disposal methods? How is solid waste handled in my community? How long does it take for materials to deteriorate? What can be done about the solid-waste problem?
Project Water Watch	Collecting data Communicating Ordering Categorizing Inferring Applying	Patterns of Change Systems and Interactions Evolution Energy Interdependence	Collect data on the physical and chemical characteristics of a river. Collect and identify macroinvertebrates of the river. Analyze the physical, chemical, and biological qualities of the river. Learn how to take action to protect bodies of water.	Where does our river come from? Where does it go? What are some indicators of water quality (physical, chemical, biological)? What can pH, dissolved oxygen levels, and the types of macroinvertebrates tell us about a river's quality? How can we protect bodies of water?
Project Soil	Categorizing Inferring Measuring Applying Communicating	Systems and Interactions Patterns of Change Interdependence Stability	Investigate the nature of soil and how it is formed. Use a variety of methods to monitor and describe soil. Collect, categorize, and analyze soil in the local environment. Compare soil in different regions of the world.	What is soil? How is soil formed? What are the major characteristics of soil? How can soil be monitored? How can soil be protected?
Project Earth Month	Categorizing Relating Inferring Applying Communicating	Systems and Interactions Patterns of Change Energy Scale and Structure Interdependence	Identify an environmental problem or issue, and design an action-taking project to help resolve the issue. Participate in action-taking projects designed to focus on environmental issues and problems.	What are some environmental problems and issues that are important to our community? How can we find out about these issues? What can we do to help resolve these problems?

The Global Thinking Learning Model

We believe that knowledge is not passively received but actively constructed by students. Therefore, this curriculum consists of a series of engaging projects that involve students in monitoring important physical and biological aspects of their environment and in collaborating within their classroom and with students in other classrooms as well. As you begin to use these materials, you will notice that each project is organized as a learning cycle and contains suggested activities for eliciting students' prior experiences and knowledge (stage 1), exploring concepts about an environmental topic (stage 2), proposing explanations and solutions (stage 3), and applying their knowledge through action-taking projects (stage 4). These four stages—invitation, exploration, explanation, and action— define the constructivist learning model we have used in the development of the projects in this book.

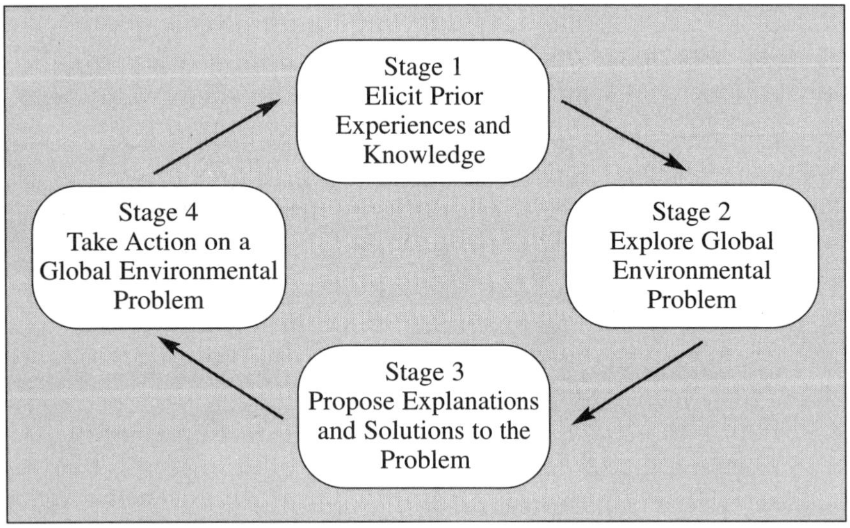

Global Thinking Learning Cycle

Establishing a Global Thinking Classroom

The materials in this resource are designed to help you establish a global thinking environment in your classroom. Many of you will be integrating the activities in this guide into your science or social studies courses. Others will be using GTP in after-school science clubs or Saturday science programs. However you choose to incorporate the projects into your curriculum, the experience will have a profound impact on the way your students think about people in other cultures, the environment, and the creative use of technology for the betterment of humankind.

A Community of Practice

The GTP links schools together using the Internet. Each classroom should have a computer, modem, printer, and link to the Internet. This technology provides a way to link students in distant places to collaborate on the study of global problems. To facilitate the concept of "community of practice," the project uses a variety of communication tools, including e-mail, bulletin boards, chat rooms and video-conferences.

How to Facilitate Cooperative Learning

The ultimate solutions to most of the problems that students will explore in the GTP will require cooperation among people and governments from different nations. The deterioration of ozone in the stratosphere, destruction of trees and other plants by acid precipitation and ozone, and the gradual rise in the Earth's temperature are problems that have multinational causes, and will require global solutions. Students who are involved in cooperative learning[3] activities develop the collaborative strategies and skills needed to do this kind of work.

Mixed ability teams comprised of two to four students work best and can facilitate work not only when students are doing activities in the classroom or collecting data outside, but when they use the computer as well. Assigning a role to each student is one way to promote interdependence among group members while helping each student develop personal responsibility for some aspect of the activity. Rotating the roles within a group from day to day helps to develop a sense of shared leadership and teamwork among group members. Here is a synopsis of roles we suggest:

Communicator

The communicator is responsible for asking the teacher or another team's communicator for help if the team gets stuck. For example, if the team cannot resolve a question or decide how to follow a procedure, the communicator is the only team member who can leave the team and ask the teacher for help or talk with the communicator of another team. The communicator shares with other teammates any information obtained from another communicator or the teacher. The communicator is the computer Internet operator for the team, and is responsible for keyboarding team messages and informing the team of messages received from other teams.

Tracker

The tracker tracks the team's progress through the steps of a Global Thinking activity. The tracker might check off each step of the procedures as the team completes each step, or remind the team members to read the step again if they are moving too quickly to the next step. The tracker is responsible for recording information for activities that require only a single team record and should work directly with the communicator when preparing e-mail messages to other schools or working on Internet activities.

Checker

The checker makes sure the team understands and completes the task. When the team begins a Global Thinking activity, the checker makes sure everyone understands the activity and helps the team discuss the task before beginning. During an activity, the checker should facilitate and encourage talk about the activity. In general, checkers facilitate the team's actions, but should not be construed as team leaders.

Materials Manager

The materials manager is responsible for picking up and returning all supplies and equipment needed for an activity. Materials managers also check that equipment, especially monitoring tools, are in good working order; if something is damaged or broken, they inform the teacher. Materials managers facilitate cleanup, but all members should participate in cleaning up!

---

[3] Johnson, D. W., and R. J. Johnson. *Learning Together and Alone,* 4th ed. (Boston: Allyn and Bacon, 1994); Slavin, R. E. *Cooperative Learning. Theory, Research and Practice* (Englewood Cliffs, NJ: Prentice Hall, 1990).

Cooperative Learning Team Role Cards

Make a copy of these cards for each team. Cut up the cards and distribute them to the designated member of each team. Cards can be folded and set on the team's table to remind students of their roles.

Communicator

- Helps resolve problems
- Can leave the team to communicate with teacher or other communicators
- Transmits messages on the telecommunications system; informs team of messages from other teams

Tracker

- Helps track progress
- Records data/information for the team
- Collaborates with communicator on e-mail and conference messages

Checker

- Helps team understand the activity
- Facilitates talk about the activity
- Is not the leader; is a facilitator

Materials Manager

- Picks up/returns materials
- Facilitates cleanup
- Checks to make sure equipment is in working order

Role of the Teacher

The Global Thinking teacher is a facilitator of learning. During the projects, facilitating cooperative learning among the teams in your class and helping them become linked with student teams in other schools will be your principal roles. You will also help your students ask testable questions, work with them to identify suitable resources for investigations, help them take responsibility for their own learning, and provide them with access to the Internet so that they can interact with global partners and find resources on the Web to help with their inquiries.

Assessing Student Learning

Anything students do can be used for assessment purposes. Active assessments such as class brainstorming, student journals and reports, e-mail, questionnaires, and concept maps allow teachers to find out more about what and how students are thinking than traditional paper-and-pencil tests. Portfolios provide information about student learning over time. We encourage you to explore these and other forms of authentic assessment as you document the nature and extent of your students' learning.

Students' learning through the GTP takes many forms. During field tests of the GTP, teachers have noted that students learned actual physical skills ("how pollution can be measured," "how to make observations about our local study site"), as well as higher-order thinking skills related to experimental design and analysis ("designing scientific experiments to find out about the environment") and problem solving ("everyone has a part in protecting the environment"). But the most important learning, according to the teachers we've heard from, came from using the Internet to connect with other students around the world. This contact resulted in an awareness of student attitudes in different countries, such as "Students in other countries are environmentally aware and concerned" and "Kids are very much alike in different countries", as well as general respect for different views and excitement about the potential of the Internet for fostering international communication and collaboration.

CHAPTER 1

Using the Internet to Collaborate Globally

From Space I saw Earth—indescribably beautiful with the scars of national boundaries gone.
Muhammad Ahmad Faris
Syrian Astronaut

One of your roles as a teacher in the Global Thinking Project (GTP) is to create a classroom environment that is conducive to communication among your students and students not only in other classes in your country but in classes around the world. A way to do this is to establish a telecommunications hub in your room that will enable your students to link with their peers. With a computer and a connection to the Internet, your students will be connected globally. This chapter will explain how to best use the Internet and its tools to achieve your goals.

A Global Classroom

Sara Crim and Galina Manke are science teachers at Chattanooga Valley Middle School, Flintstone, Georgia, and Experimental Gymnasium School 710, Moscow, Russia, respectively. Their classrooms are each equipped with a computer, a modem connected to a telephone line, and a printer. Their schools have had telecommunications connections to the Internet since 1990, enabling them to connect with schools not only in their own country, but all around the globe. They were two of the first teachers who participated in the GTP.

By using the Internet, you will be able to create a global classroom that will link your students with classrooms all over the world.

Like these teachers, you and your students will be participating in the Global Thinking Project. One of your first responsibilities is to establish a global classroom. To do so, you should have the following in your room (more specifics on these later).

- Microcomputer
- Connection to the Internet via a dial-up system or a direct connection[1]
- Modem
- Printer
- Internet software
  - E-mail
  - Web browser
  - Telnet
  - FTP (File transfer protocol)
- VCR
- CD-rom
- Projection system (so you can show the Web and e-mail to the whole class)
- Large map of the world
- Globes

Ideally you should strive to have one computer for every four or five students, thereby enabling each of the cooperative teams to have its own Internet connection to the tools presented in this chapter. If, however, you have access to only one computer, establish an Internet center in your classroom. Set up the computer and a printer in the Internet center on a round table that can fit three or four students around it. Place a nearby bulletin board containing a map of the world with adequate space to post messages retrieved from your telecommunications system. Make available computer magazines, journals, books, and general resources on environmental science and computers. Start collecting software related to global and environmental problems and make it available to your students.

Taking the time to establish an Internet center will be well worth the effort. Students will know that because you established the center, they are able to connect with peers around the world. And that is a compelling thought.

[1] Dial-up connections to the Internet require the use of a modem, whereas direct connections use a connection to a router to the Internet.

Getting Online

Being connected to the Internet will enable you to use e-mail and to make use of browsing tools such as Netscape Navigator® and Internet Explorer® to share and retrieve data and search the Internet for information about the environmental topics and issues in the projects.

Here is what you will need to get started:

1. **Telecommunications Software.** Telecommunications software enables your modem and computer to talk to each other and then interact with the Internet. You may have telecommunications software in your school's computer and not actually know it. For example, ClarisWorks® has a telecommunications component that you could use. Many Internet Service Providers (ISP) either give or sell (typically at low cost) specially designed software enabling you to connect to the Internet.
2. **A Computer.** It really doesn't matter what computer platform you use, Macintosh® or IBM® PC. You should have either a Mac with at least 32 megs of RAM, or a PC with Windows®. However, if you want to run software with easy-to-use graphical screens, then older computers such as an Apple IIe® or IBM® 286 won't work. But if you have an Apple IIe® or an IBM® 286, then you can still connect to the Internet.
3. **A Modem—for dial-up systems.** A modem is a device that converts signals that computers use into a form that can be sent over telephone lines. The word *modem* is short for *mod*ulator/*dem*odulator. Purchase a fast speed fax/modem, one that is at least 28.8 bps (bits per second). Higher speed modems will enable you to use graphic software and download image files from the Internet in relatively short periods of time.
4. **A Telephone Line.** A regular telephone line will work. If there is none in your classroom currently, the best choice is to have one installed. If there is a phone jack near your classroom (perhaps in the science department office), it is possible to share the line by running a line off of that line into your classroom. If you are going to set up a telecommunications system at home, you might want to consider a separate fax/data telephone line. If not, you can use your current line.
5. **An Internet Provider.** You need a phone number that your modem dials to connect you to an organization that provides access to the Internet. First, check with your county or district Office of Education and ask what options or plans there are for Internet connections. In some states, such as Florida (FIRN), Georgia (PeachNet & EduNet), and Texas (Tenet), Internet access is provided via a statewide network. There are a number of commercial and not-for-profit

telecommunication providers. (See Resources.) You might want to consider Mindspring, NETCOM, or the Institute for Global Communications (IGC). Each of these providers has Internet accessible software, and each supports science education projects. (See Resources.) You might also check with a local university or college and ask if they provide Internet access to local teachers.

6. **Optional Tools.** Identify someone in your school who has been using telecommunications and seek out their support and help. In most cases, they will be more than willing to help. Many teachers who use the Internet with their students have discovered that having accessibility to the Internet at home increases their success in using telecommunications in the classroom. If you can't get a phone line into your classroom at first, get started at home. You can use your home computer to do a few start-up projects with your class, or use it to learn about the Internet and begin to make professional contacts with other science teachers. This is the way some of the early GTP teachers began.

One more thing. Your students are a powerful resource when it comes to the Internet. Survey them and find out how many use the Internet regularly. These students can be an invaluable resource for using the World Wide Web, developing Web pages, and engaging in other Internet activities.

GTP Internet Technology

The Internet tools that are discussed below are all integrated into the GTP Web site. You will find it useful to understand the types of technology used in the GTP to create an interactive virtual community. As you start using the GTP Web site, each of the tools will be used for a different function. In each of the cases described below, a software program enables you to achieve a type of interactivity.

E-mail

The GTP began as an e-mail project. This is the simplest form of electronic communication and is available to nearly anyone that joins the GTP. Older computers and schools that do not have fast connections to the Internet can still use e-mail and, therefore, participate in many of the GTP activities. You can use any type of e-mail software. Some of the popular programs used by GTP teachers are Eudora, ClarisWorks®, and Pine.

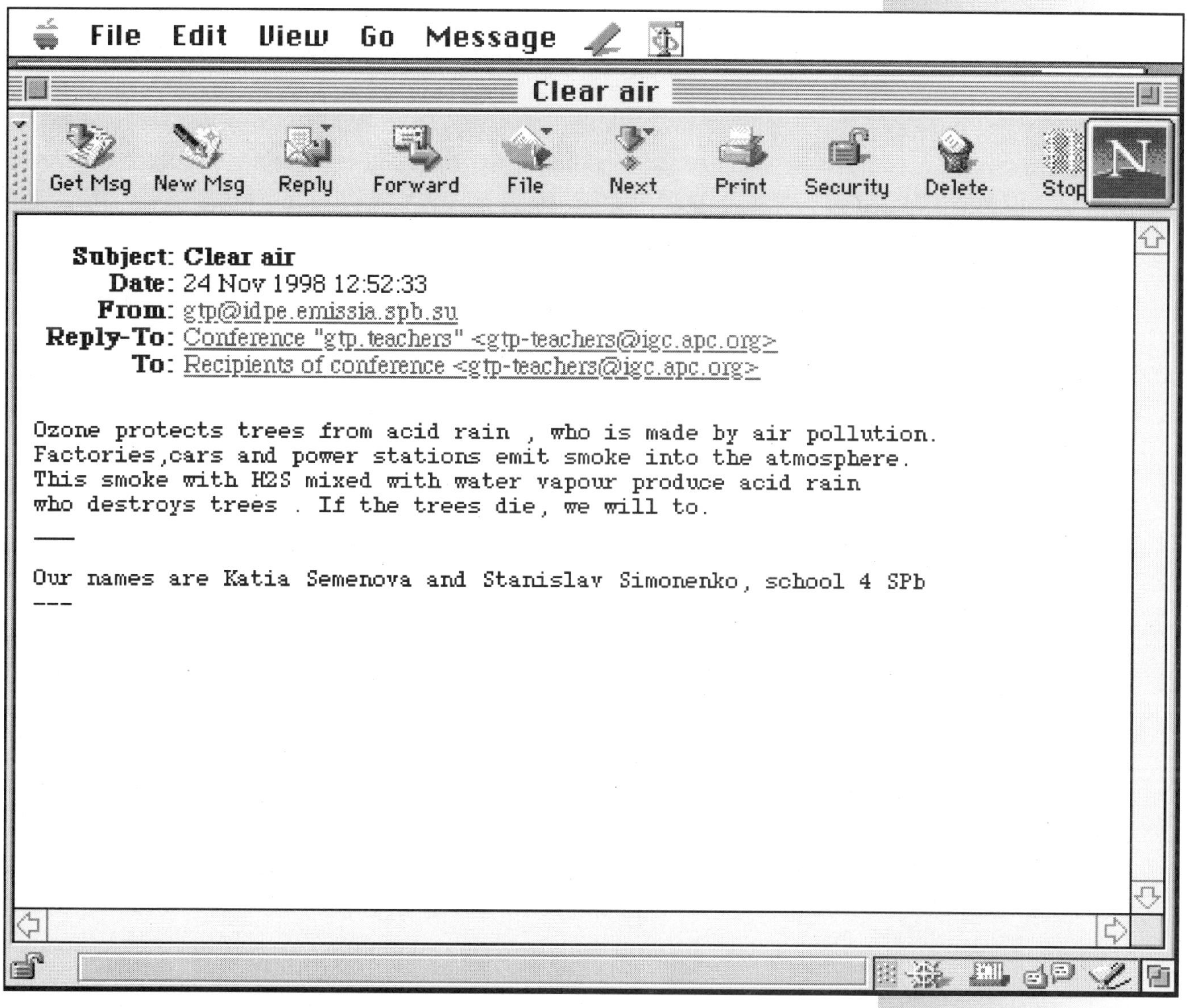

Subject: Clear air
Date: 24 Nov 1998 12:52:33
From: gtp@idpe.emissia.spb.su
Reply-To: Conference "gtp.teachers" <gtp-teachers@igc.apc.org>
To: Recipients of conference <gtp-teachers@igc.apc.org>

Ozone protects trees from acid rain , who is made by air pollution.
Factories,cars and power stations emit smoke into the atmosphere.
This smoke with H2S mixed with water vapour produce acid rain
who destroys trees . If the trees die, we will to.

Our names are Katia Semenova and Stanislav Simonenko, school 4 SPb

---

An e-mail message from one GTP school to another.

E-mail messages should always be signed by you or the students who authored the message. In many e-mail programs you can set up an automatic signature that is printed at the end of a message. There are three things you should keep in mind when sending e-mail messages to GTP schools:

1. For many of the readers of your messages, English is not the primary language. Keep your messages brief, and be aware of the complexity of words and phrases that you use in composing your messages.

2. Remind your students that they are trying to build relationships among teachers and students from around the world. Part of this relationship-building is responding to e-mail. Try not to delay in replying to e-mail that you receive. To accomplish this, you might assign groups of students to be responsible for e-mail correspondence on a daily or weekly basis.
3. Bring to your students' attention some of the ethical issues related to Internet use.

Internet Ethics List[2]

1. Thou shalt not use a computer to harm other people.
2. Thou shalt not interfere with other people's computer work.
3. Thou shalt not snoop around other people's files.
4. Thou shalt not use a computer to steal.
5. Thou shalt not use a computer to bear false witness.
6. Thou shalt not use or copy software for which you have not paid.
7. Thou shalt not use other people's computer resources without authorization.
8. Thou shalt not appropriate other people's intellectual output.
9. Thou shalt think about the social consequences of the program you write.
10. Thou shalt use a computer in ways that show consideration and respect.

Listserves

The GTP uses a listserve to enable you to send e-mail that is distributed to everyone participating in the GTP. The nice thing about the GTP listserve is that you only have to type in one address to reach everyone. When you join the GTP, you will be prompted to subscribe to the GTP listserve, thereby putting you and your students in touch with teachers and students in other schools. You can use the GTP to communicate efficiently with everyone and keep in touch with GTP staff.

Conferences: Electronic Bulletin Boards

On the Internet, a conference is a form of an electronic bulletin board. You send a message to the "conference," and it is placed there for others to read. But conference software also will enable you to organize e-mail messages into a logical set of categories. Within the GTP Web site, we have established a conference for each of the GTP projects. These conferences are actually online discussions that will enable your students to confer with other students, topic by topic. The online discussions are easily accessed. They appear on the Web page for each specific GTP project.

---

2 Developed by the Computer Ethics Institute. Reprinted by permission.

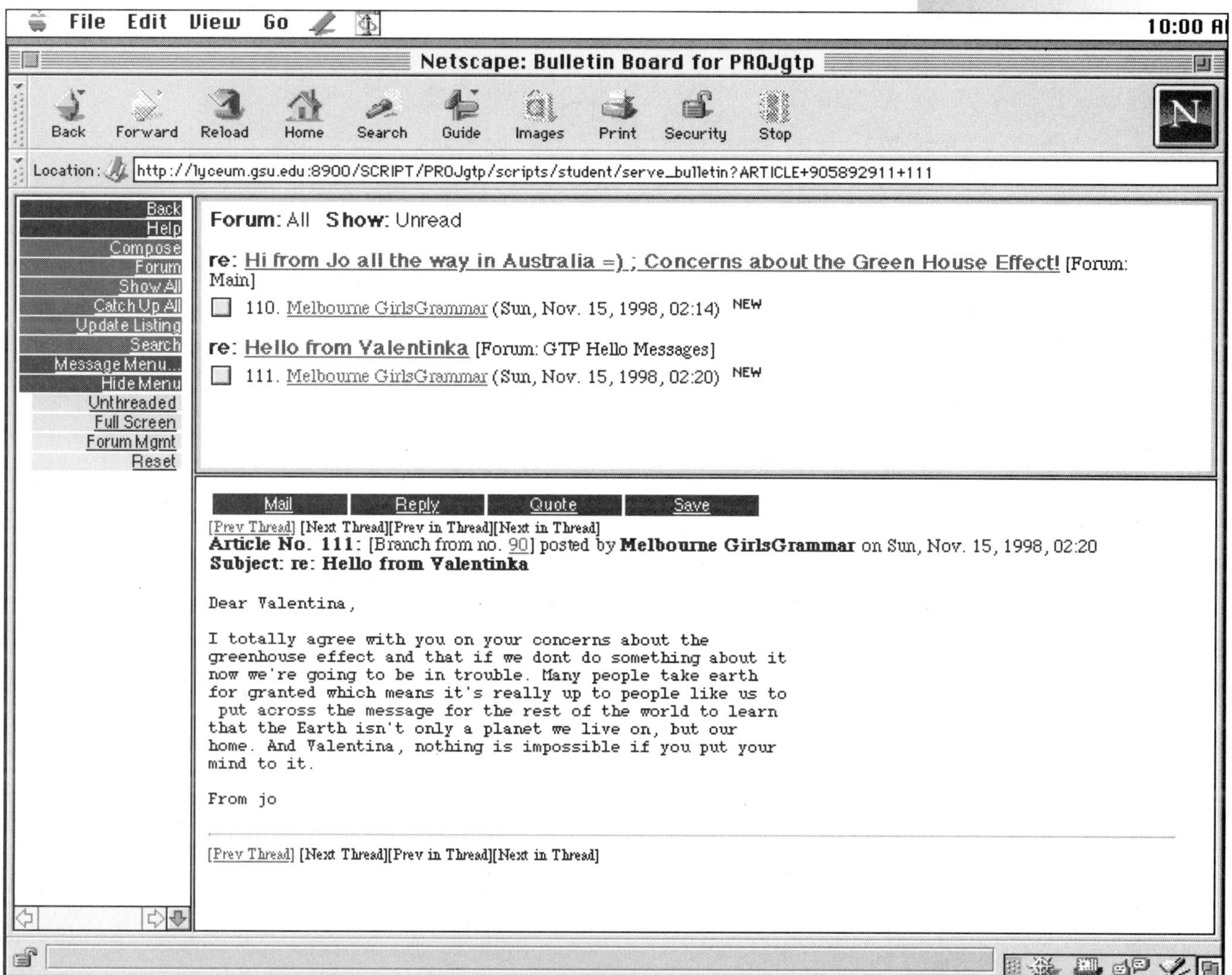

A sample message sent to the GTP Bulletin Board. Students can send messages to the Bulletin Board, and read messages sent by others in the GTP.

Chat

Chat requires "real time" connections among GTPers. We have established a "chat room" where GTP participants can discuss GTP activities and projects. Chat requires that an announcement be made in advance stating that a real-time chat will occur on a specific date and time. To announce a chat, all you have to do is post a message on the GTP listserve.

Chat sessions are limited to a specific number of participants. Typically ten to fifteen sites may participate in a chat session. If you plan to conduct a chat session, send a message to the GTP community using

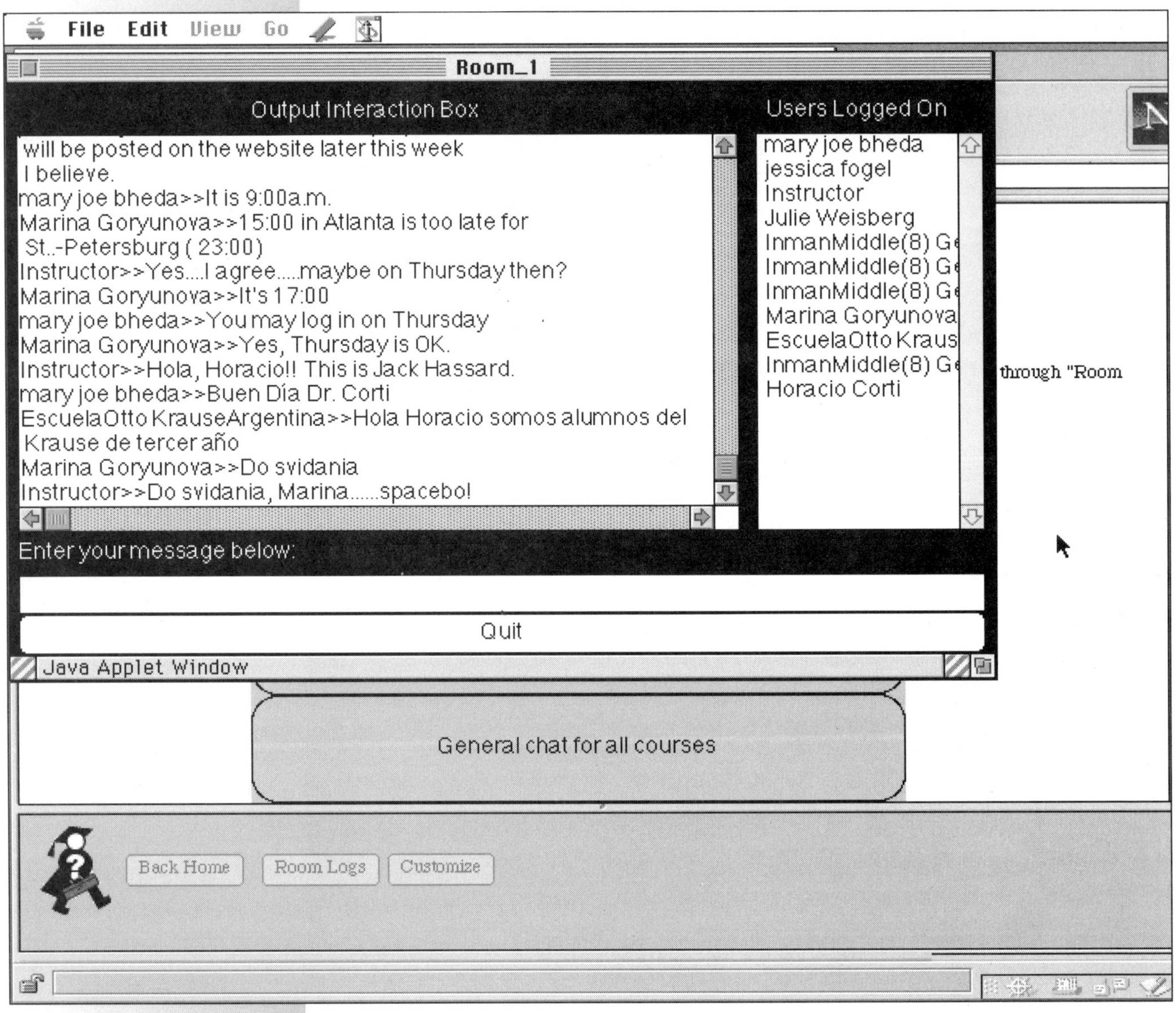

A GTP Chat Window
Students and teachers from several GTP schools can chat about GTP research in "real-time."

the GTP listserve, announcing what the topic will be, the date and the time, and approximately how long it will last. Typically chat sessions should be no more than an hour.

There are a number of considerations to keep in mind when chatting on-line. The most important is an awareness of the time zones involved in the GTP. For example, the chat sessions held between American students on the East coast and Russian students in the Moscow region usually take place around 9 A.M. Eastern Standard Time, which means the Russian students log on at about 4 P.M. Moscow time.

Video Conferences

Using inexpensive digital cameras (e.g., the Connectix® black and white or color cameras) and desktop video-conferencing software (e.g., CU-SeeMe®), GTP students and teachers in different locations can communicate by sound and sight. Use this technology to participate in point-to-point and group conferences.

The GTP has set up a reflector site that is used only by GTP participating schools to arrange for and carry out video conferences. Information about the GTP reflector site (specifically its Internet address) can be found by visiting the GTP Web site. The reflector enables you to participate in a group video conference. For example, you might want to originate a conference on Project Hello after initial e-mail messages have

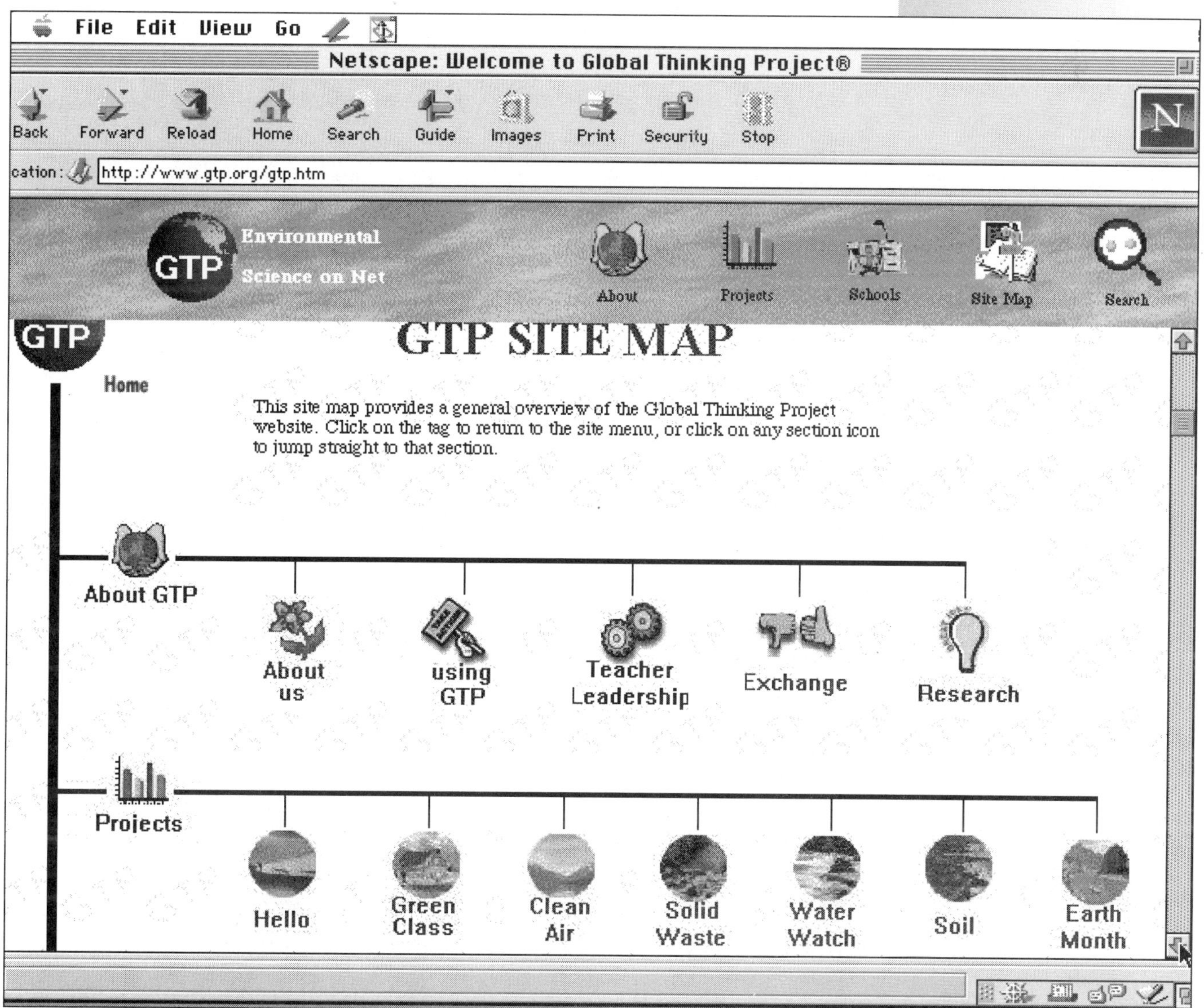

GTP Web Site Map

IP (Internet Protocol) Address:
an address assigned to your computer when connected to the Internet. It usually has four numbers separated by periods, such as 144.188.311.6.

been exchanged. The reflector provides a collaborative environment where your students can participate in a conference with students from other GTP schools.

It will also be possible to arrange for point-to-point conferences between your school and another GTP school. To do this, you will need the *IP address* of the other school, which can be obtained by contacting the school. This is much like simply dialing up a friend to make a phone call. Once you start collaborating with other schools using e-mail, you might find it fun to try and arrange for a two-way conference.

The Web and the GTP

The Web serves as an important technology for you. By using the GTP Web site, you will become a participant in a virtual global community bringing together teachers and students from around the globe. The GTP Web site has been designed so that you can carry out the activities and projects of the GTP online, as well as to enable you to quickly and easily access information on each of the GTP projects.

To access the Web, you will need a browser—a software program that allows you to read pages created on computers that use a special language named HTML (Hyper Text Markup Language). Netscape Navigator® and Microsoft's Internet Explorer® are two of the most commonly used Web browsers. The browser will be one of the most important tools for you and your students. With it you will be able to not only "surf the net," but also to participate in interactive activities within the GTP Web site. You might want to stop what you are doing right now, and look at the GTP Web site. The Web site map on page 9 shows how information is organized within the Web site.

Summary of Internet Technologies

The Internet technologies that have been discussed can be divided into two categories: interpersonal tools and informational tools. Internet tools can help us collaborate with others on the Internet, as well as find and publish information on the Web. Let's take a closer look at these two categories.

Interpersonal Tools

The Internet was initially established as an interpersonal medium by which researchers could easily and efficiently share information on the research they were conducting. As the Internet grew, new "tools" were developed to enrich the way people were able to communicate with each other. As you work with the GTP, think of some Internet technologies as "interpersonal tools," enabling you and your students to do things such as send e-mail letters around the world to other GTP classes, to participate in online discussions of important environmental problems, and even to participate in real-time chats. And on some occasions, you might want to participate in GTP video conferences.

Informational Tools

The Internet is a network of networks connecting computers and people. Although it is hard to keep up with the size of the Internet, in the late 1990s there were millions of people connected through the Internet.[3] This number continues to expand by leaps and bounds. Furthermore, it was estimated that there were at least five million computers on the Internet that were "serving up" information. The Internet includes text, graphics, video, and sound, and it is organized using "hypermedia" documents. In the early 1990s, a tool was invented that enabled Internet users to browse the computers on the Internet that were serving information. The original browser, called Mosaic, led to the development of other browsers, such as Netscape Navigator® and Microsoft's Internet Explorer®.

When you visit the GTP Web site you will be able to access information about each of the GTP projects, as well as participate in chat, video, and other communication activities. You will also be able to find out who is participating in the GTP, and then use this information to contact other teachers to carry out collaborative projects. You will also find Web pages for each of the GTP projects (such as Project Clean Air, Project Water Watch, Project Soil, and Project EarthMonth). These Web pages will serve as organizing centers for your class as your students become engaged in the various GTP projects.

Web-Assisted Teaching

Web-Assisted Teaching: using the resources of the Internet as part of your lesson plans

The GTP provides a framework for using a variety of teaching resources and tools to engage your students in environmental studies projects. By using the GTP materials, you will be implementing a Web-assisted teaching program. The resources include hands-on materials for use by students to monitor environmental factors inside and outside the classroom, lesson plans that you can use to implement environmental science activities in a cross-cultural context, and computer tools, including a student data analysis program and software for using the Web.

[3] Williams, B. *The World Wide Web for Teachers.* Foster City, CA: IDG Books Worldwide, 1996.

Web-assisted teaching means that the resources of the Internet are part of your teaching lesson plans. The Web is utilized to help you achieve a number of learning goals that you would not normally be able to attain. These goals include, but are not limited to:

- Interactivity and collaboration with other schools around the world
- Participation in data-sharing activities whereby your students share and receive data and ideas with students from other schools
- Computer file searches all over the world to research environmental science topics
- Participation in a global community of students and educators interested in environmental issues and project-based teaching and learning
- E-mail reports and publishing Web pages that can include text, pictures, drawings, audio, and video about environmental issues

Teaching with the Web as a resource opens up the possibility of designing learning activities that would not have been possible before the Internet. For example, although many teachers have used pen pals as a medium to foster collaboration among students around the world, postal mail is slow, and often does not meet the goals of our learning activities. On the other hand, by using the powerful tool of electronic mail, students are able to participate in a range of "key pal" activities (the electronic equivalent of pen pals).

The GTP utilizes a number of Internet activity types, and they are clustered into three categories:

- **Interpersonal Exchange Activities:** Activities that focus primarily on personal communication among students, teachers, and others such as scientists, politicians, activists, and citizens
- **Information Collection Activities:** The exchange and organization of data and other information that is either generated by students or obtained from other sources, such as databases, books, journals, and magazines
- **Problem Solving Projects:** Activities that combine interpersonal exchange activities and information collections to foster student inquiry and problem solving

Interpersonal Exchanges

In the GTP, the Internet enables users to communicate in key pal exchange, as a member of a global classroom, or with experts.

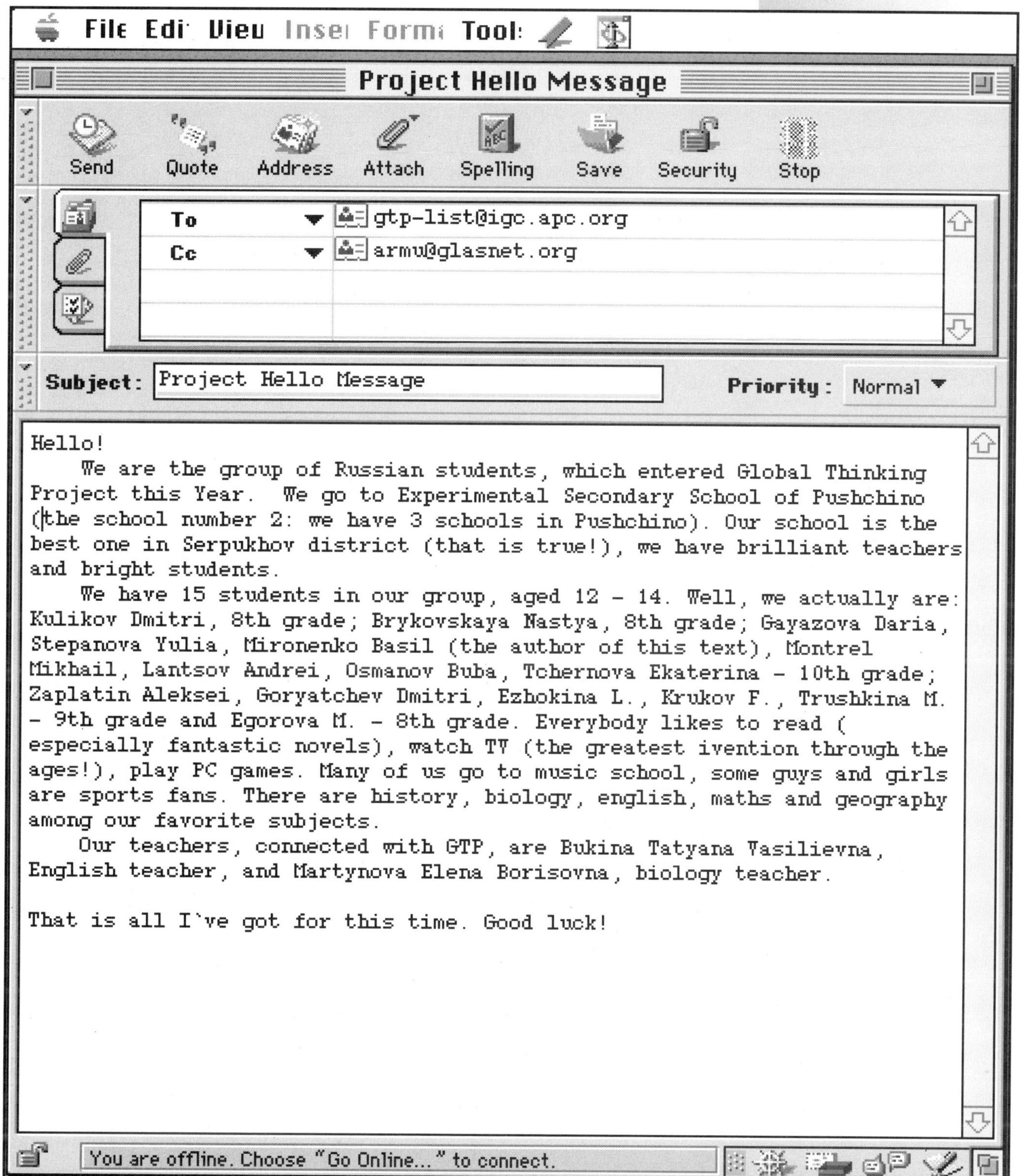

Project Hello Message

To: gtp-list@igc.apc.org

Cc: armu@glasnet.org

Subject: Project Hello Message

Priority: Normal

Hello!

We are the group of Russian students, which entered Global Thinking Project this Year. We go to Experimental Secondary School of Pushchino (the school number 2: we have 3 schools in Pushchino). Our school is the best one in Serpukhov district (that is true!), we have brilliant teachers and bright students.

We have 15 students in our group, aged 12 – 14. Well, we actually are: Kulikov Dmitri, 8th grade; Brykovskaya Nastya, 8th grade; Gayazova Daria, Stepanova Yulia, Mironenko Basil (the author of this text), Montrel Mikhail, Lantsov Andrei, Osmanov Buba, Tchernova Ekaterina – 10th grade; Zaplatin Aleksei, Goryatchev Dmitri, Ezhokina L., Krukov F., Trushkina M. – 9th grade and Egorova M. – 8th grade. Everybody likes to read (especially fantastic novels), watch TV (the greatest ivention through the ages!), play PC games. Many of us go to music school, some guys and girls are sports fans. There are history, biology, english, maths and geography among our favorite subjects.

Our teachers, connected with GTP, are Bukina Tatyana Vasilievna, English teacher, and Martynova Elena Borisovna, biology teacher.

That is all I`ve got for this time. Good luck!

Sample Hello E-mail Letter

Key Pals. Key pals enable you to design activities in which individual or small cooperative groups of students send e-mail to each other. Project Hello starts this process by having your students send brief messages to other schools in the GTP.

Global Classrooms. Most teachers participating in the GTP use their class as a unit for collaborating with others in GTP activities. The global classroom structure will involve all of your students working together as a learning group as they complete GTP activities, log on to the Internet to send and receive e-mail, and surf the Web. Global classrooms have no boundaries. You will use the Internet as a medium to link your students with others using the GTP around the world.

Question and Answer Services. Your students will be involved in several environmental science projects in which they will conduct research to construct ideas and answer questions about the environment. Questions will arise about a topic and the Internet can be used by your class to solicit answers from "experts." There are a number of "question and answer" services on the Internet that you can use.

Information Collection

Information Exchanges. In the GTP, your students will be exchanging information on the quality of the air, water, and soil in your local environment with other GTP classes. The GTP information exchanges engage your students in the monitoring and development of an understanding of your local environment. This information is then shared with others so that they can compare it with their data. The exchange of information can be accomplished using e-mail, chat, video conferences, or Web data forms.

Database Creation. You will probably want to organize the information that students collect on the local environment. One way to do this is to set up a database. We think it is important to organize the information you receive from other schools as a valuable record of your students' work. Many GTP teachers have also organized the e-mail that was sent and received into a notebook.

There are several database programs that you can use such as ClarisWorks®, which has a database management program built in. Students can create forms using the database and then store the information. Another program you might use is Alice, which you can download free from a site at TERC (teaparty.terc.edu/tech /tech.html). With Alice you can create data tables, and then use graph and map tools to analyze and draw conclusions about GTP data. You can set up a data table for each of the GTP projects such as the one shown below.
The data tables you create with Alice will be a convenient way to organize information collected by your students. We suggest creating data tables for the following projects:

- Project Green Classroom (Chapter 3)
- Project Clean Air (Chapter 4)
- Project Solid Waste (Chapter 5)
- Project Water Watch (Chapter 6)
- Project Soil (Chapter 7)

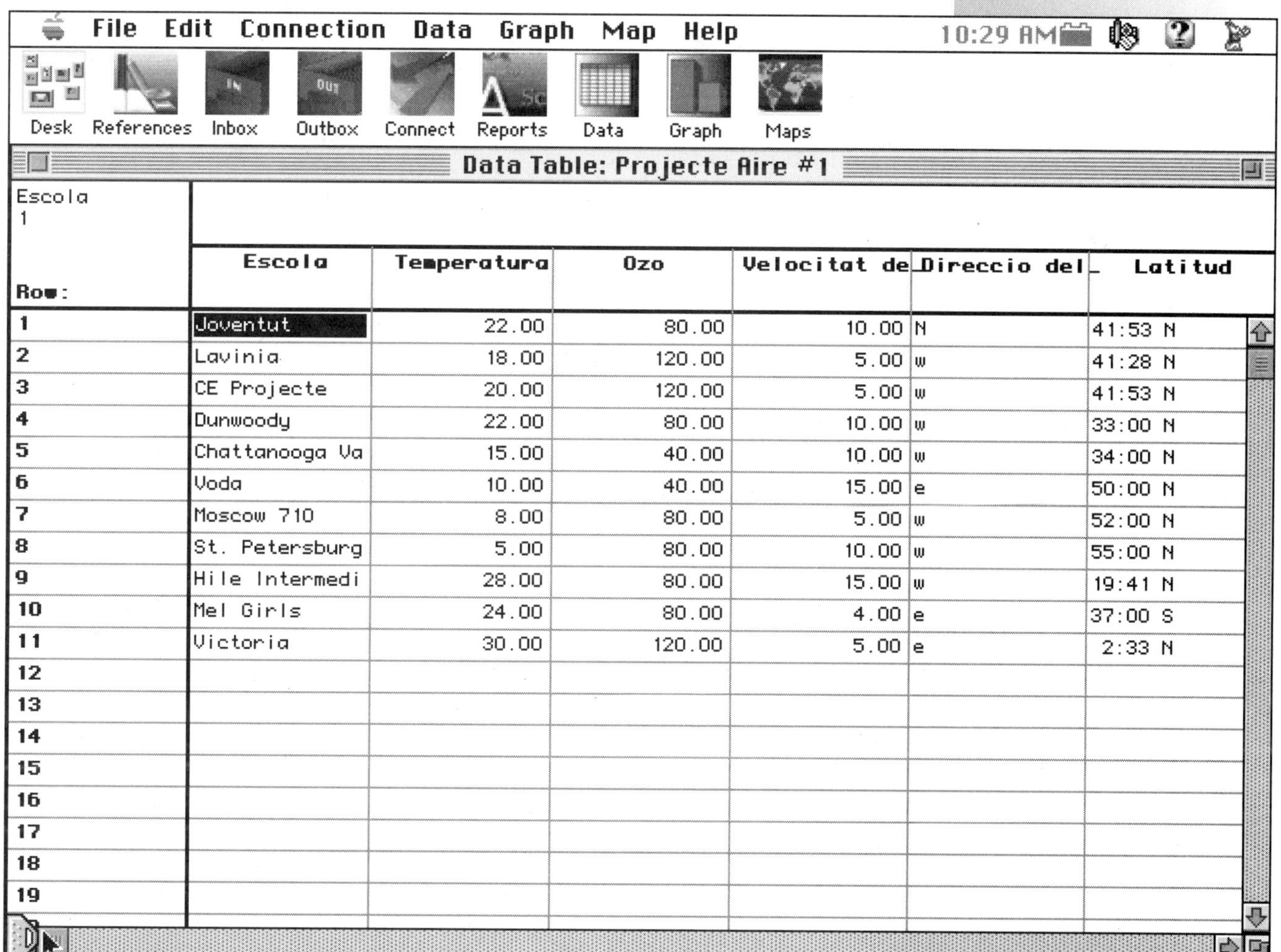

Row:	Escola	Temperatura	Ozo	Velocitat de	Direccio del	Latitud
1	Joventut	22.00	80.00	10.00	N	41:53 N
2	Lavinia	18.00	120.00	5.00	w	41:28 N
3	CE Projecte	20.00	120.00	5.00	w	41:53 N
4	Dunwoody	22.00	80.00	10.00	w	33:00 N
5	Chattanooga Va	15.00	40.00	10.00	w	34:00 N
6	Voda	10.00	40.00	15.00	e	50:00 N
7	Moscow 710	8.00	80.00	5.00	w	52:00 N
8	St. Petersburg	5.00	80.00	10.00	w	55:00 N
9	Hile Intermedi	28.00	80.00	15.00	w	19:41 N
10	Mel Girls	24.00	80.00	4.00	e	37:00 S
11	Victoria	30.00	120.00	5.00	e	2:33 N
12						
13						
14						
15						
16						
17						
18						
19						

Alice Data Table

It is also possible to retrieve data directly from the GTP Web site. To do this, go to the home page for the GTP Project containing the data that you wish to retrieve. From the home page point to the "data retrieval" link. When you open this page, you will see a table of data that you can download directly to your computer. Follow the online instructions to do this. When the data is downloaded, it will appear on your hard drive as a file that can be opened by Alice or by your favorite spreadsheet program.

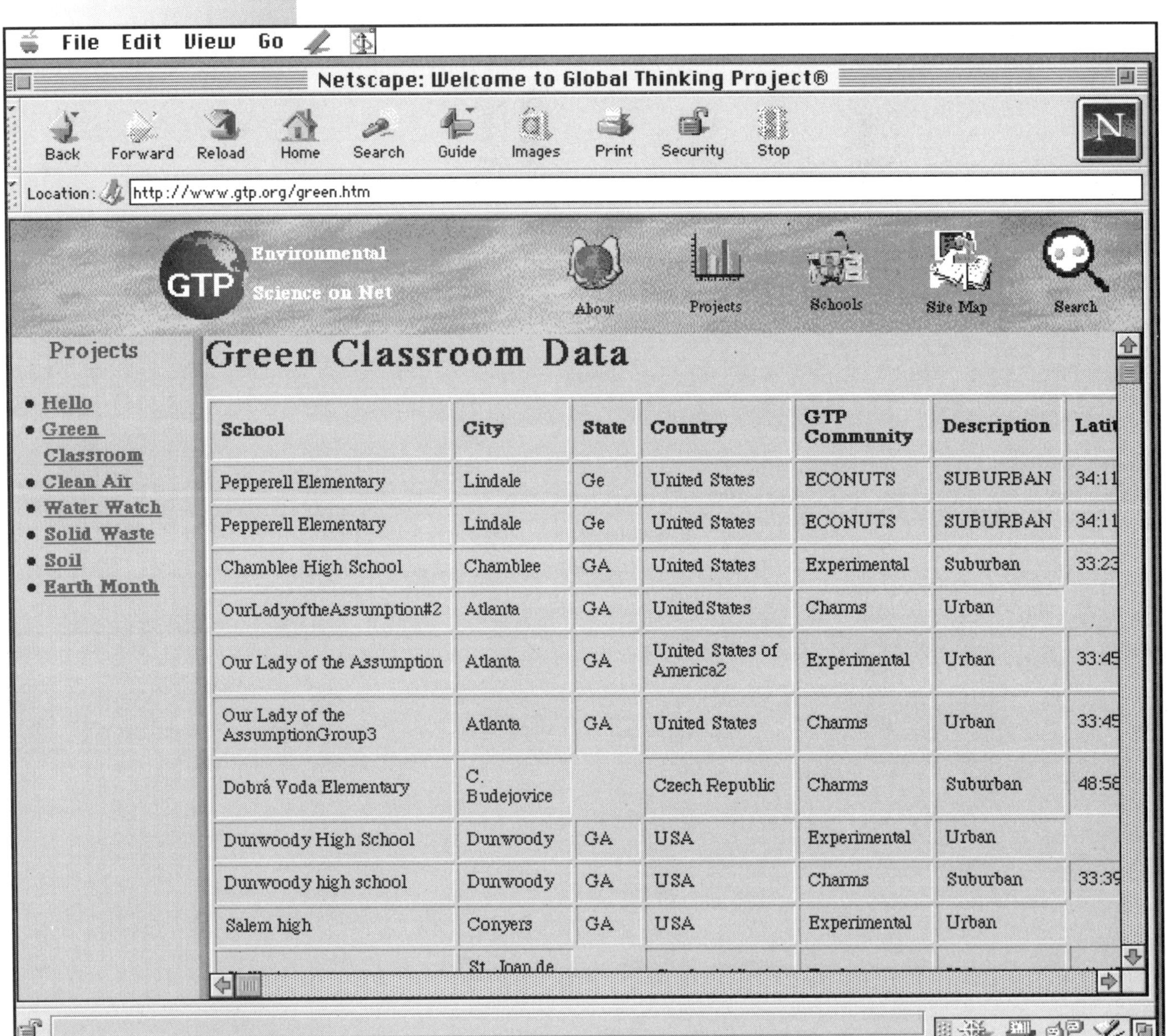

School	City	State	Country	GTP Community	Description	Lati
Pepperell Elementary	Lindale	Ge	United States	ECONUTS	SUBURBAN	34:11
Pepperell Elementary	Lindale	Ge	United States	ECONUTS	SUBURBAN	34:11
Chamblee High School	Chamblee	GA	United States	Experimental	Suburban	33:23
OurLadyoftheAssumption#2	Atlanta	GA	United States	Charms	Urban	
Our Lady of the Assumption	Atlanta	GA	United States of America2	Experimental	Urban	33:45
Our Lady of the AssumptionGroup3	Atlanta	GA	United States	Charms	Urban	33:45
Dobrá Voda Elementary	C. Budejovice		Czech Republic	Charms	Suburban	48:58
Dunwoody High School	Dunwoody	GA	USA	Experimental	Urban	
Dunwoody high school	Dunwoody	GA	USA	Charms	Suburban	33:39
Salem high	Conyers	GA	USA	Experimental	Urban	
	St. Joan de					

Project Green Classroom Data Table
After data is reported to the GTP, it can be retrieved as a data table and downloaded to your computer.

Pooled Data Analysis: Network Science. Pooled data analysis is scientific inquiry that utilizes the Internet to enable students to "pool" data they collect locally for comparison with data collected by students from around the world. This type of analysis fosters collaboration and it also focuses on a series of important problem solving steps.

Pooled data analysis in the GTP includes using Internet forms to share data, retrieving GTP data posted by participating schools, and using graphing and mapping tools to display the data in a variety of forms. Sending and sharing data that your students have collected is relatively easy. From the home page of one of the GTP Projects, point to the "data sharing" link. A form such as the one shown below will appear. Enter data

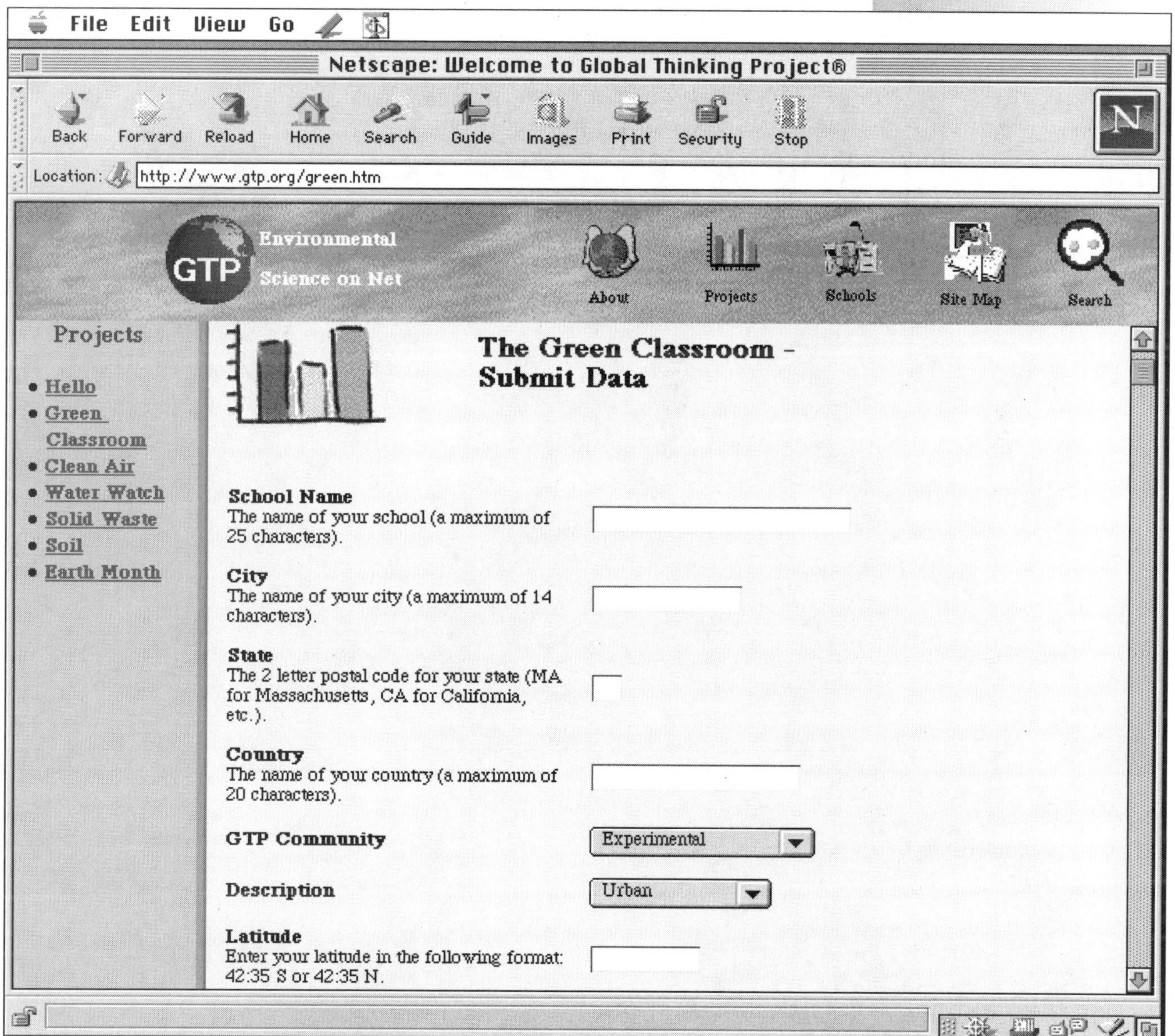

Project Green Classroom Data Entry Form
Internet forms share data with the entire project group.

directly on the form and send to another computer which organizes the information and places it into a data table that can be read and retrieved.

Pooled data analysis also includes graphic analysis of the pooled data. Your students can make graphs, charts, and maps of the data using Alice or some other spreadsheet program, such as ClarisWorks®. The graph below was generated from GTP clean air data, and the map view on page 19 was generated using the ozone data taken from the same table.

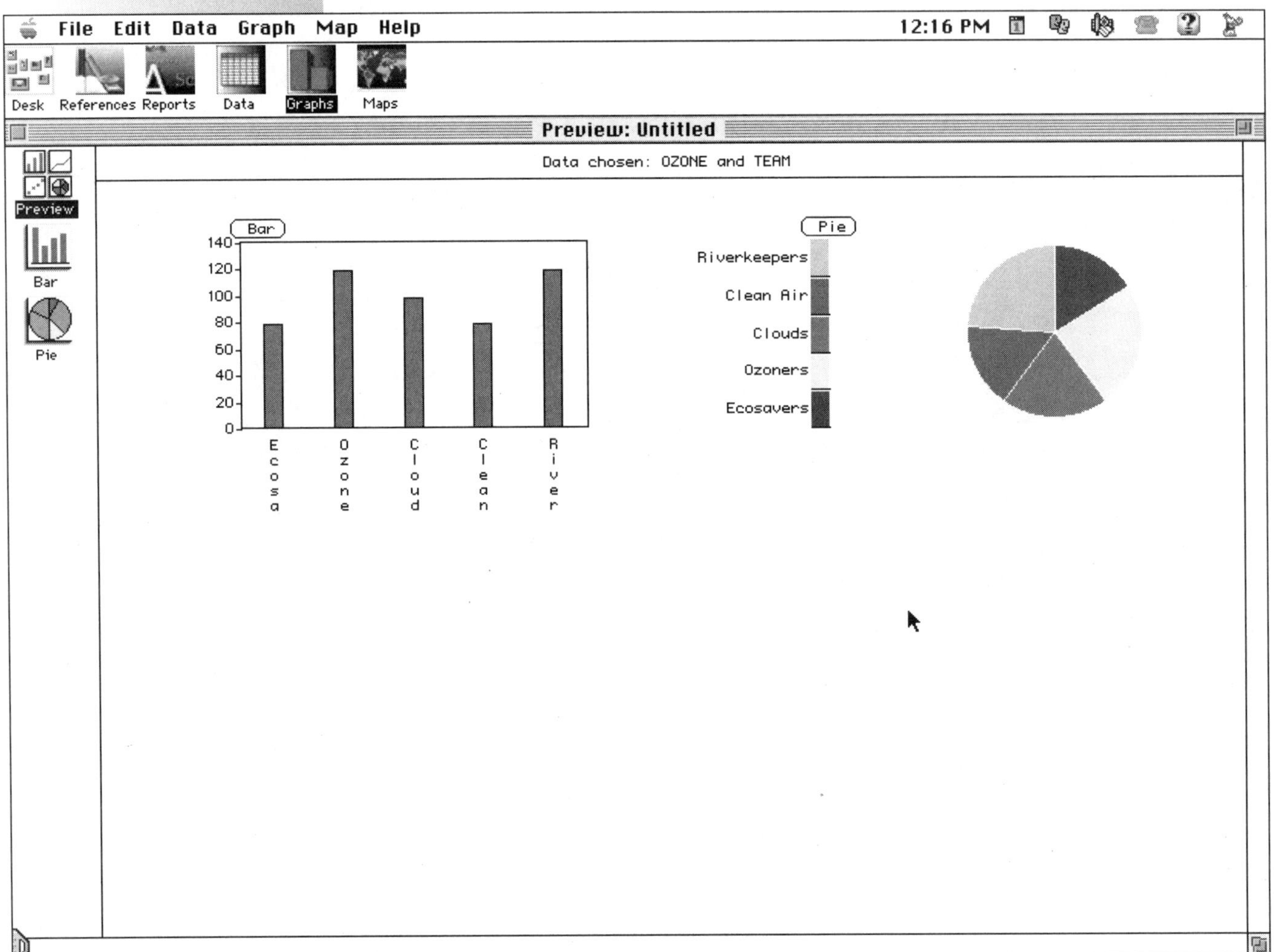

Graph of Ozone
Ozone vs. temperature graph, created using Alice.

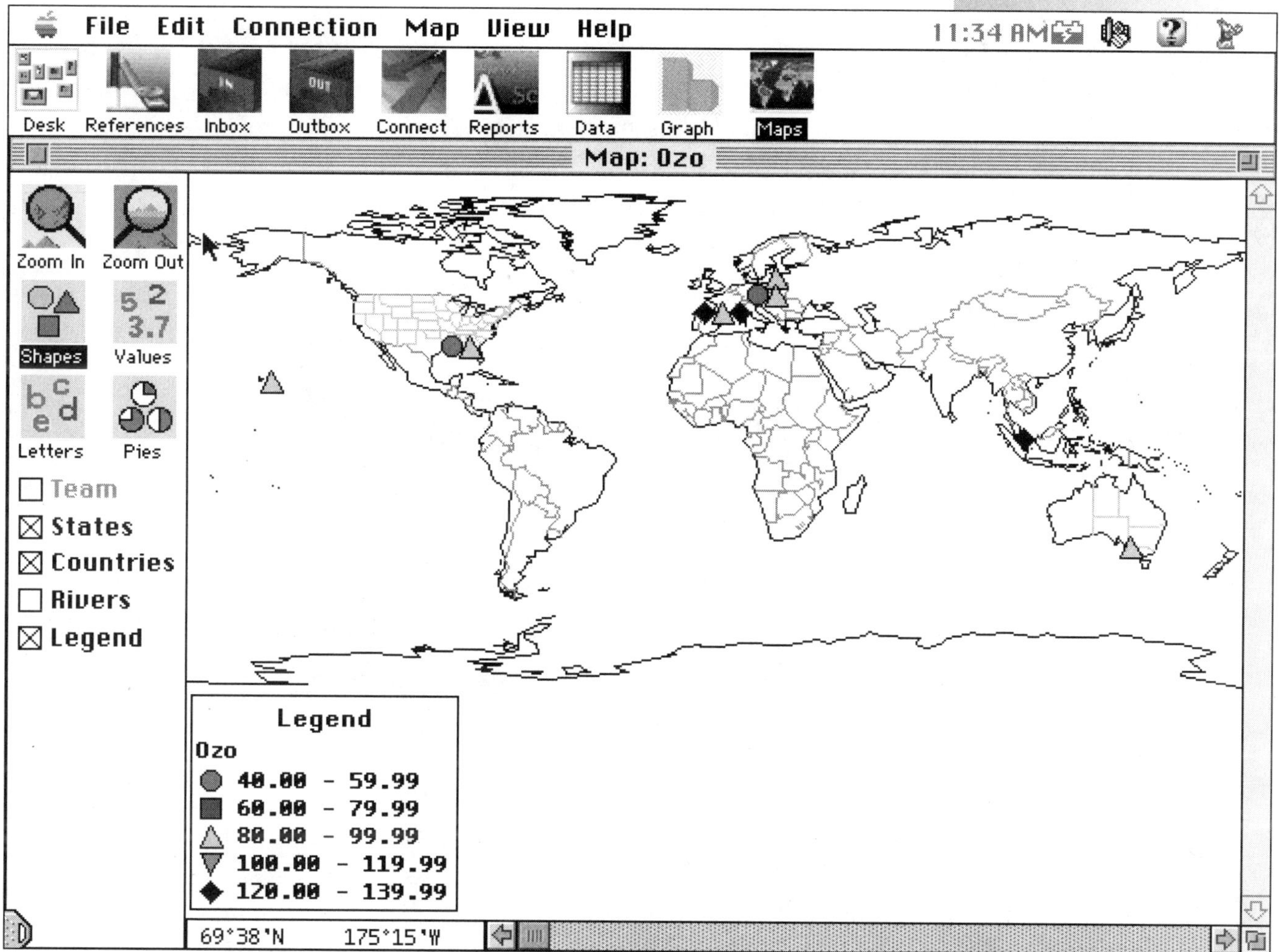

Map of Ozone Concentrations
This map was drawn using Alice. The data is depicted by using symbols, each of which represents a different value.

Problem Solving Projects

All of the GTP projects focus on problem solving by involving students in information searches, data collection, analysis, and social action. To make your work efficient, we have organized each of the projects as a chapter in this book, and have developed a home page for each project that you can access from the GTP Web site. For example, the Web page below shows the home page for Project Clean Air (Chapter 4). If you go to this Web page, use the GTP lesson plans in Chapter 4, and provide each student with a copy of the GTP Student Journal for Project Clean Air, you will have the major teaching tools needed to engage your students in a GTP problem solving project.

You will find detailed lesson plans for each project in this book, including learning goals, hands-on materials, monitoring procedures, and specifics on how to easily use the Internet as an integral part of the GTP.

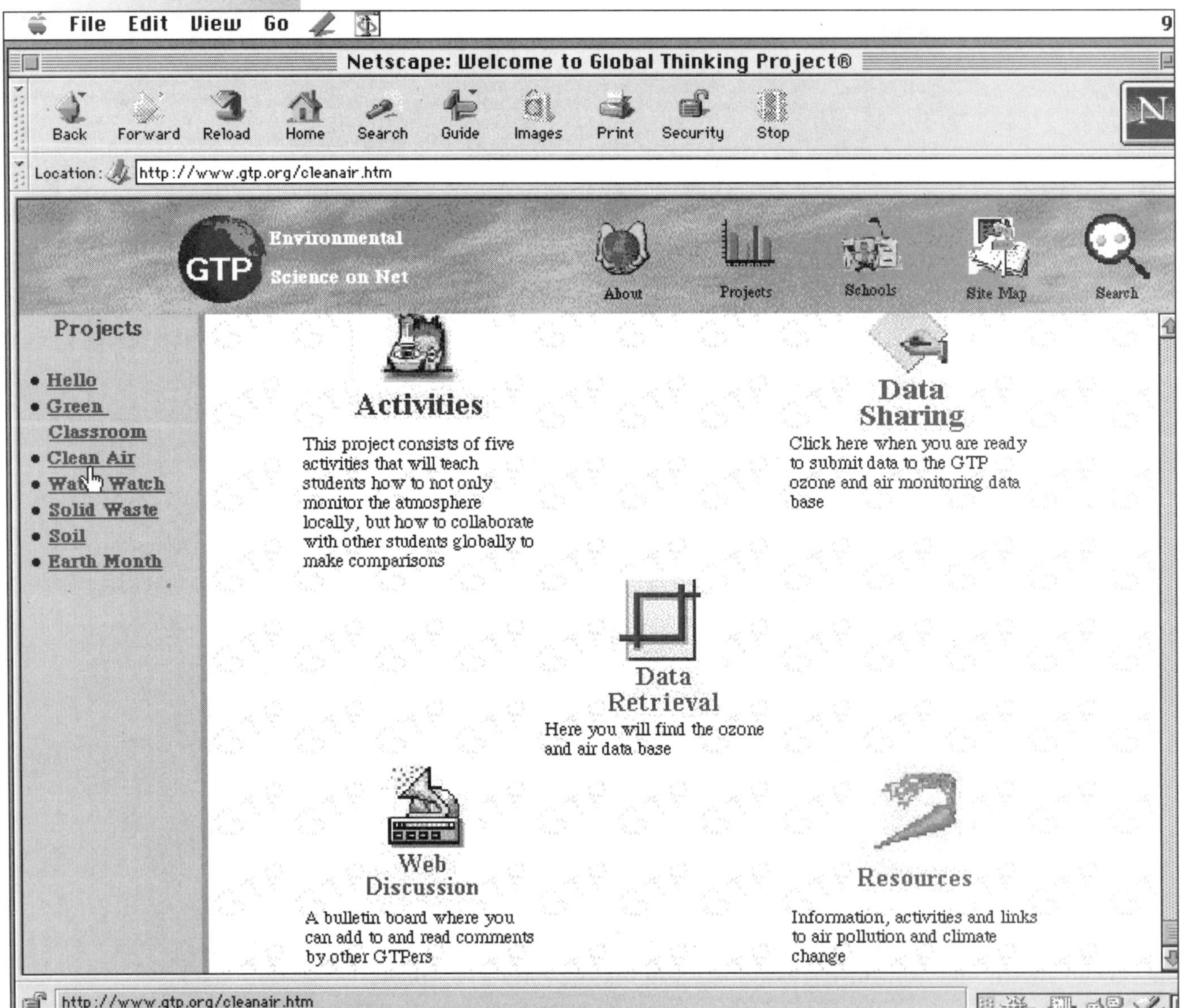

Project Clean Air Home Page

CHAPTER 2

Project Hello

Global thinking is thinking how our actions and reactions affect all the world.
Nadia, age 13

Interconnectedness is one of the central concepts underlying the GTP. Helping students around the world understand their role and their stake in maintaining the fragile equilibrium among Earth's resources is the goal of the GTP. In Project Hello, students explore their initial ideas about Global Thinking and global environmental problems, and learn how to use the Global Thinking Web site to share data and discuss findings with students in other schools.

The planet Earth, as seen from Apollo 17, December 1972.

Two "mini-projects" included in the unit are the creation of a Global Thinking bulletin board and the exchange of Hello Pacs. Many teachers have found that a bulletin board is a good place to post data and e-mail messages, as well as articles and information related to student projects. Students have also enjoyed exchanging special packages, called Hello Pacs (containing pictures, articles, newspapers, and artifacts about themselves and their school) with other Global Thinking classes.

Planning Alert

Make Earth Systems card sets prior to Activity 2.3. (See page 29 for instructions.)

Photocopy sets of Global Problems cards prior to Activity 2.4. (See page 32.)

Goals for Project Hello

1. Students will explore their initial ideas about Global Thinking.
2. Students will become familiar with the GTP Web site.

Before beginning this project, you will need to divide your class into teams of four and to devise a means of assigning and rotating the cooperative learning roles among team members. Use the following planning chart as a guide.

Activity	Class Periods
2.1 What is Global Thinking?	1
2.2 Earth as a Living System	1–2
2.3 Global Environmental Problems	1–2

Planning Chart

Web Site Information

From GTP home page (www.gtp.org) click on “projects” to access the Project Hello home page.

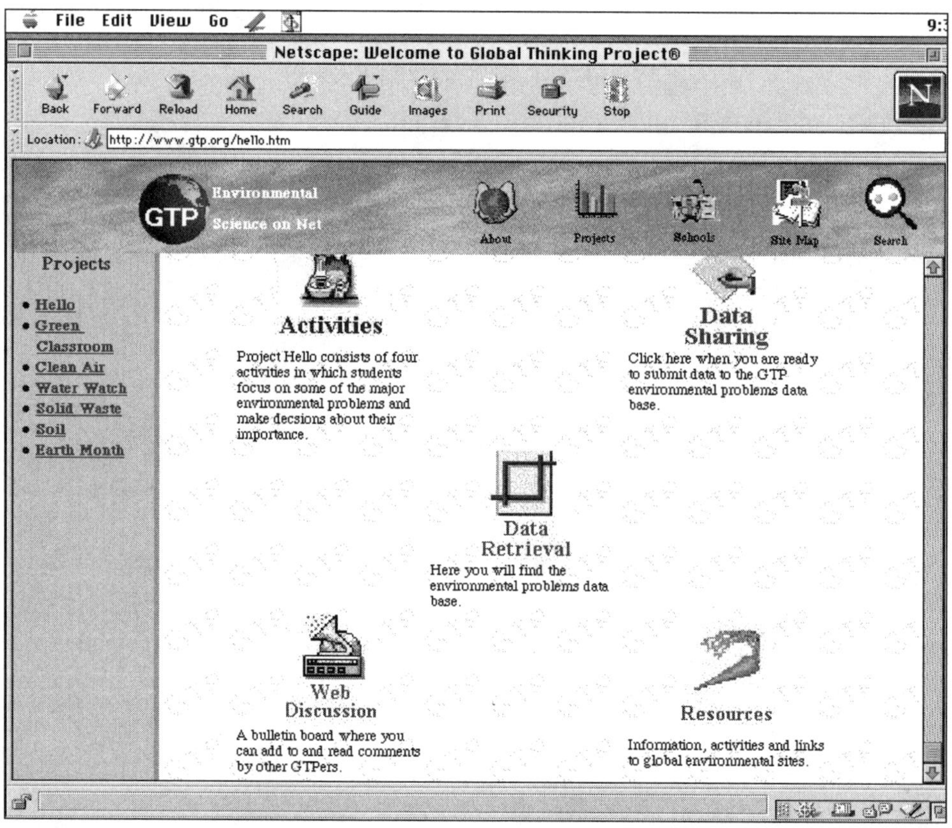

Project Hello Home Page

ACTIVITY 2.1

What Is Global Thinking?

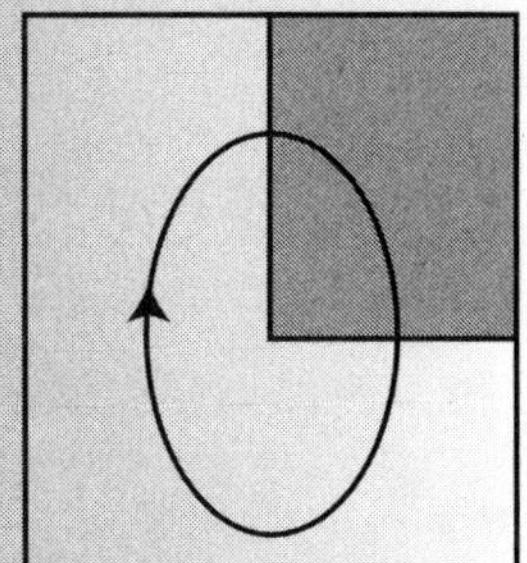
Invitation

In this activity, students are introduced to the GTP, and explore their initial notions of what the term *global thinking* means. They formulate an operational definition of global thinking that will be subject to modification and refinement throughout the remainder of this project.

Objectives

- Describe the nature of the GTP.
- Define operationally the term *global thinking*.
- Locate schools participating in the GTP on a world map.

Materials

world map, sticky dots or pins, chart paper, marking pens, Student Log pages 187, 188

Procedures

1. Explain to students that they are going to be involved in the GTP, and that they will be studying local environmental problems and issues and collaborating with students in other states and countries through a WWW home page and e-mail.

   Demonstrate for the students how to go to the GTP home page using your Internet browser, and use the resources available through the home page to identify the countries involved in the project this year. Help the students locate these countries on a world map and mark them with sticky dots or pins.

   Open the Project Hello home page. Acquaint the students with the resources and databases available through this home page. Later in this project, students will add and retrieve data from the Project Hello database.
2. Tell the class that the focus of their work will be global thinking. Show the students a picture of Earth, such as the one at the beginning of this chapter. Then read this quote by Yuri Artyukhin, a Russian cosmonaut, who said this after one of his voyages in space.

   It isn't important in which sea or lake you observe a slick of pollution, or in the forests of which country a fire breaks out, or on which continent a hurricane arises. You are standing guard over the whole of our Earth.

Take a few minutes to discuss this quote with the students. Can they paraphrase what Artyukhin is saying? How would orbiting Earth contribute to such thinking?

3. Give each team the task of brainstorming their answers to two questions: "What, in your opinion, is global thinking?" and "What questions do you have about global thinking?" Tell the students to write at least five different ideas for each question on a diagram like the one shown below on a sheet of chart paper. Then ask them to summarize their ideas into a single statement and to identify the most important question from their list. Give them ten minutes to complete this task. (Remind students to use the cooperative learning roles: materials manager, recorder, tracker, communicator.)

Global Thinking T-chart

What, in your opinion, is global thinking?	What questions do you have about global thinking?
Summary of statements above:	**Most important question from above:**

4. When each team is finished, the communicator should write the team's results on a large piece of chart paper in front of the room. When all have finished, read and discuss the results. Invite students to add any concepts or questions they feel have been omitted. Hang the statements and questions on a wall in the classroom so they can be referred to during the remainder of this project.

Class Mini Project #1

Create a Global Thinking Bulletin Board

To acknowledge the importance of global thinking in your class, create a Global Thinking Bulletin Board in conjunction with your students. Many classes create interactive bulletin boards in their classrooms that contain information relevant to the study of global thinking, including the names and locations of participating schools, data from the most recent class investigations, and recent e-mail and conference messages. You might form a bulletin board committee comprised of one member of each team, or devise a system for rotating the responsibility for maintaining and updating the bulletin board from team to team.

Ask the students what they think should be included in such a bulletin board. They will suggest a variety of items, such as a world map, newspaper and magazine articles about global problems, and information about collaborating schools. As students bring in articles from magazines and local newspapers, post them on the Global Thinking Bulletin Board. It is important to keep this information current. You might assign groups of students the responsibility for keeping this section of the bulletin board up-to-date.

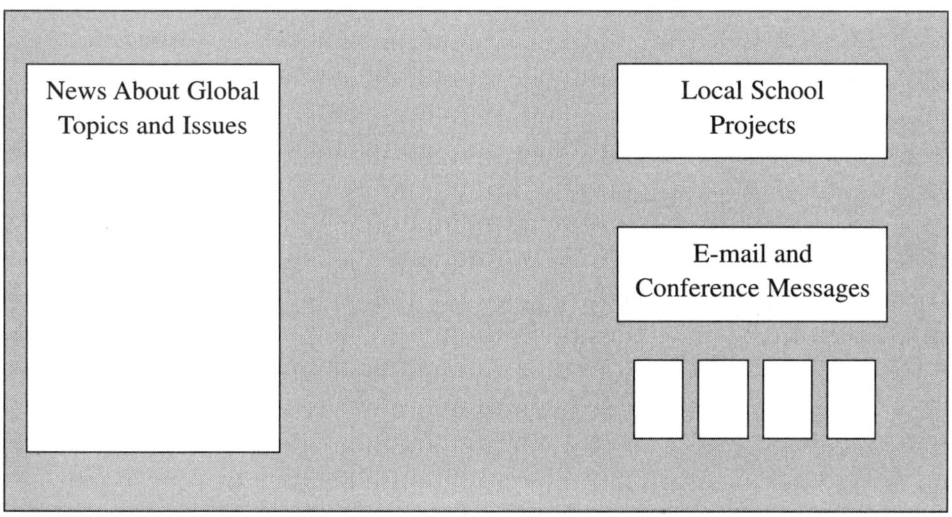

Global Thinking Bulletin Board

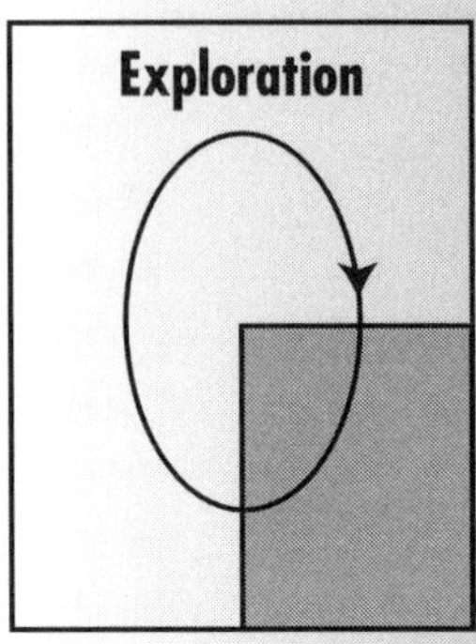

ACTIVITY 2.2

Using the Internet to Think Globally

The purpose of this activity is to introduce the Global Thinking Project Web site to students, and to discuss how it will be used as a tool to collaborate with other schools participating in the GTP. Teams will write and post their first messages to the Project Hello bulletin board.

Objectives

- Become familiar with the GTP and Project Hello home pages.
- Introduce students to other schools participating in Project Hello by posting a message on the Project Hello bulletin board.

GTP Home Page

Materials

computer, modem connected to telephone line, Web browser, printer, glossary pages from this resource, Student Log page 189

Procedures

1. Point out to students the computer setup in your classroom. Ask your students a series of questions to find out what their prior knowledge and experiences are with computers and telecommunications. For example, you might ask: "How many of you . . .

   - have access to a computer at home?"
   - have taken a computer course?"
   - have used the computer to send e-mail?"
   - can explain the purpose of a modem?"
   - have played computer games?"
   - have 'surfed the Net'?"

2. Demonstrate how your computer connects to the Internet. Show students how to use your Web browser to open the GTP home page (www.gtp.org). Show students how they will be able to navigate to projects, to find out about the GTP and current news, to find out what schools are participating, and to search the Internet for information.
3. **Communicator Orientation**—Meet separately with the communicators from all of the teams to make sure they know how to open the GTP home page and are familiar with the GTP navigation bar and the GTP Web site map. Each communicator will be responsible for making sure all of his/her team members have mastered this skill. Review Internet terms from the glossary with these students, so that they can explain these terms to their teammates.

3a. **"Hello" Messages**—While you are working with the team communicators, the remaining team members should compose brief "hello" messages that will introduce their team to the other schools participating in Project Hello. Have the students use Student Log page 189 to compose drafts of their e-mails before writing and sending from the Web site. In addition to personal information about team members (names, ages, families, grade in school), they might also want to include some information about their school or community. Encourage students to include some of their initial ideas about global thinking (developed in Activity 2.1) and to ask for comments and feedback from others.

This is a good time to remind students about e-mail etiquette. Emphasize that a good e-mail letter is short and to the point, and that it uses simple language and short sentences so that it can be understood by non-native English speakers. Also remind the students to be sure to sign their names and e-mail addresses to their letters.

4. Sending "Hello" Messages—Allow time for each team, under the direction of its communicator, to keyboard its "hello" message, and post it to the Project Hello bulletin board.

> To: Global Thinkers
> Re: Hello
>
> ______________________________
>
> ______________________________
>
> ______________________________
>
> ______________________________
>
> Signed:
> Names of students
> School
> Team name (if appropriate)
> E-mail address

Hello Message Format

Class Mini Project #2
Create and Send "Hello" Pacs

With your students, assemble "Hello" Pacs, small packages of artifacts to be sent to schools in the project via the postal mail system. Ask each team to contribute ideas for items to be included in the package. Be sure to include a photograph of the class members, with their names; a photograph of the classroom or photographs of the school; and some information about your school and/or town. Local maps and/or newspapers are also useful. Make agreements to exchange these "Hello" Pacs with classes with whom your students develop online relationships.

When you receive "Hello" Pacs from other classes, add materials from these to your Global Thinking Bulletin Board.

Exploration

ACTIVITY 2.3

Earth as a Living System

This activity introduces the concept that Earth can be viewed as a living organism (Gaia),[2] with the lithosphere, atmosphere, hydrosphere, biosphere, and heliosphere interacting to form a functioning whole. Changes that affect any one of these subsystems necessarily affect them all. Culture can be viewed as an expression of the choices people make about how they live in relationship with other forms of life and with the planet. Thus, global environmental problems are related to how people utilize resources to satisfy their wants and needs.

Objectives

- Develop a model representing the interrelationships between Earth's resources and living things.
- Reconsider the meaning of the term *global thinking.*

Materials

Earth system cards[3]: to prepare cards for the students, you will need to find pictures, or make drawings of the following elements. Create (or have students create) four large cards of 1) (heliosphere (the sun); 2) the atmosphere (or sky); the hydrosphere (oceans, river); and lithosphere (ground, earth, rocks). Prepare a set of three smaller cards consisting of 1) a group of people; 2) plants; and 3) animals. Photocopy the cards to create the number of sets you will need for your class; Student Log page 191.

Procedures

1. Using the cards, introduce the terms *heliosphere, lithosphere, atmosphere, hydrosphere,* and *biosphere.* Ask students to explain in their own words what they think these terms mean.
2. Form four large discussion groups containing one member of each team. Assign each large group one of the resources (heliosphere, lithosphere, hydrosphere, atmosphere) and ask them to brainstorm answers to the following question: "What does this resource provide that is necessary for the maintenance of the other resources and of life?" Each student should take notes on the discussion so that he/she will be able to communicate these results to his/her team. Allow about ten minutes.

---

[2] Lovelock, J.E. 1972. "Gaia as Seen Through the Atmosphere." *Atmospheric Environment 6,* 579–580.

[3] The Earth system cards used in this activity are based on the "Our Planet, Our Home" learning system developed by Dr. Philip Snow Gang. For further information, contact him at ties.tmn.com.

3. Reorganize the students into their teams. Allow ten minutes for students to share the results of their discussions with their teammates.
4. Ask each team to construct a model of the interrelationships they see between the four resources and plants, animals, and humanity, using the Earth system cards. Students should show relationships by connecting components of their system with arrows or string. When students have completed their models, take the class on a tour of the models. Ask the communicator from each team to explain the thinking behind each team's model.

   This is always an interesting moment, because students will have constructed their models in such different ways. It is also a good time for you to reinforce the concept of interdependence in this system by removing one card at a time from the system and asking students what would happen without it. (The system crashes, except when humanity is removed!)
5. As an extension of this activity, invite students to add some human needs (e.g., food, clothing, shelter, communication, transportation, security, recreation, affiliation) to their models. Discuss the stresses that satisfaction of these needs put on the system.
6. Summing Up—Allow five to ten minutes for teams to add any concepts or questions to their lists from Activity 2.1.

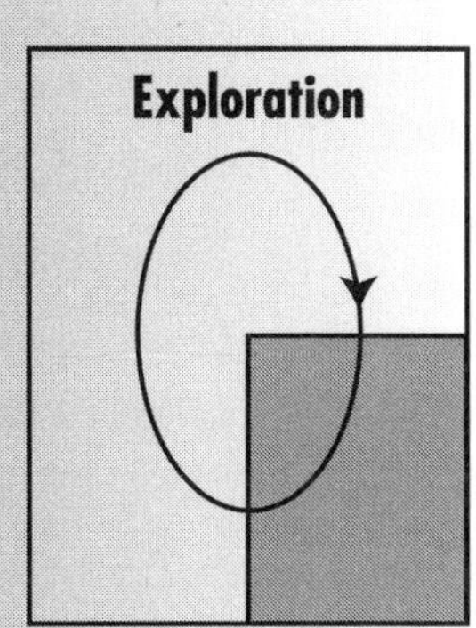

ACTIVITY 2.4

Global Environmental Problems

What are some of the problems that the planet faces today? How are these problems perceived by students in different communities and nations? The twelve problems that are identified in this activity are critical problems that need to be discussed and acted upon. Use the activity below to begin a discussion of how these problems may be perceived differently around the world.

Objectives

- Rank order global environmental problems.
- Discuss the significance of global problems and reflect upon the opinions of others.
- Compare and contrast perceptions of global problems of students from different cultures.
- Make predictions about the future of global problems.

Materials

flip chart or chalkboard, Global Problems cards (one set per team; for template, see page 32), Student Log pages 193, 194

Procedures

1. Begin by asking the class, "What are some environmental problems facing Earth today?" Write student responses on a flip chart or on the board.
2. Present the Global Problems cards, explaining that these problems have been identified in a research study as critical problems that need to be discussed and acted upon. Photocopy page 32 and cut the template into 12 individual cards. Give each team a set of cards.

Global Problems Cards Template

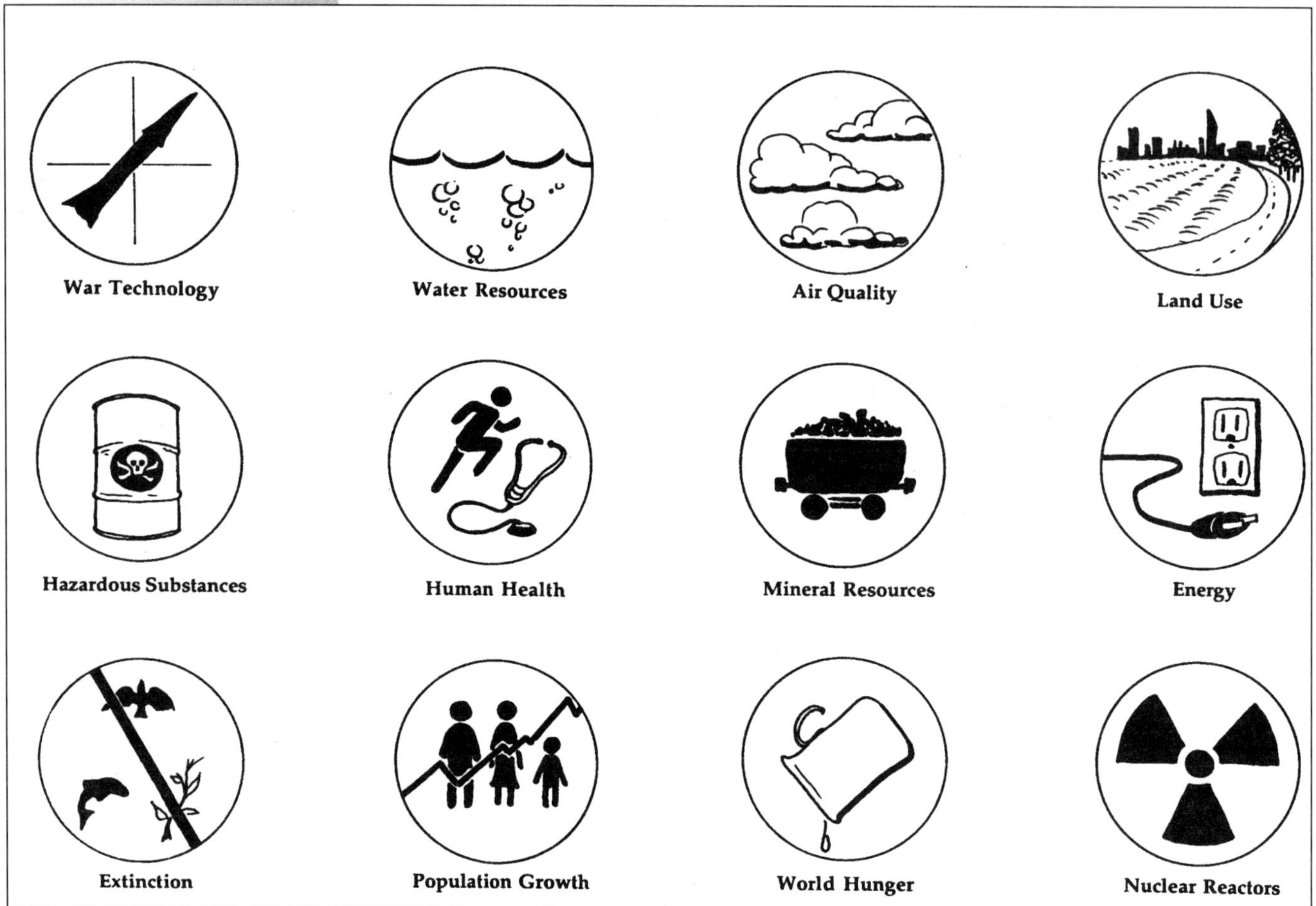

3. Allow students in teams to discuss what they know about each of the global problems represented on the cards. Then teams should rank order the problems from most to least important. Each team should reach a consensus on a ranking of the global problems. After each team has ranked the problems, they should discuss the reasons for their group's decisions. All group members should be prepared to defend their ranking. Record the team's rankings on a class data chart like the one below.

Class Data Chart

	Team 1	Team 2	Team 3	Team 4	Team 5	Average Rank
War technology						
Hazardous substances						
Extinction						
Water resources						
Human health						
Population growth						
Air quality						
Mineral resources						
World hunger						
Land use						
Energy						
Nuclear reactors						

4. Allow time for one member of each team to explain his/her team's rationale for the rankings to the rest of the class.
5. Ask students to turn to page 195 in their student logs and complete the questionnaire: How Will Global Problems Change by 2020?

   After each team member has responded individually to the questionnaire, the teams should try to reach consensus on how it thinks the problems are likely to change. Each team then should design a graph, chart, or poster and present the team's predictions to the rest of the class.

Global Problems Survey Sample

For each global environmental problem below, students indicate whether they think the problem will be better, about the same, or worse, by the year 2020.

Environmental Problem	Much Better or Better	About the Same	Worse or Much Worse	Don't Know
War Technology (nerve gas, high-tech weaponry, nuclear arms threat)				
Hazardous Substances (waste dumps, toxic chemicals, lead paint)				
Extinction of Plants and Animals (reduction of genetic diversity)				
Water Resources (waste disposal, estuaries, supply, distribution, ground water contamination, fertilizer contamination)				
Human Health (infectious and noninfectious disease, stress, diet and nutrition, exercise, mental health)				
Population Growth (world population, immigration, carrying capacity)				
Air Quality and Atmosphere (acid rain, carbon dioxide, depletion of ozone layer, global warming)				
Mineral Resources (non-fuel minerals, metallic and non-metallic minerals, mining, technology, low-grade deposits, recycling, reuse)				
World Hunger and Food Resources (food production, agriculture, cropland conservation)				
Land Use (soil erosion, reclamation, urban development, wildlife habitat loss, deforestation, desertification)				
Energy Shortages (synthetic fuels, solar power, fossil fuels, conservation, oil production)				
Nuclear Reactors (nuclear waste management, breeded reactors, cost of construction, safety)				

6. To extend this activity, use the Project Hello bulletin board to obtain rankings of global problems or predictions from students in other schools. Are there similarities or differences that emerge? Are any of these differences attributable to location, culture, age, course of study, gender, or any other variable?

   As an alternative, administer the Global Problems questionnaire to other groups in your community (students of different ages, parents, business people, scientists, etc.). How are their responses similar or different from yours?

Using the World Wide Web

Data Transmission

After the students have compiled data for Activity 2.3 and arrived at a class set of rankings for the global environmental problems, the class should submit its data to the GTP Project Hello database. From the Project Hello home page, select "Data Sharing." Students will be able to enter their data directly into a Web form and send it electronically to the GTP database.

Following is a list of the data that will be submitted:

- School name
- City
- State
- Country
- Number of students responding
- A number indicating the rank for each of the global problems, from 1 (most important) to 12 (least important)
- Results of the global problems questionnaire, reported as a percentage of respondents in each category

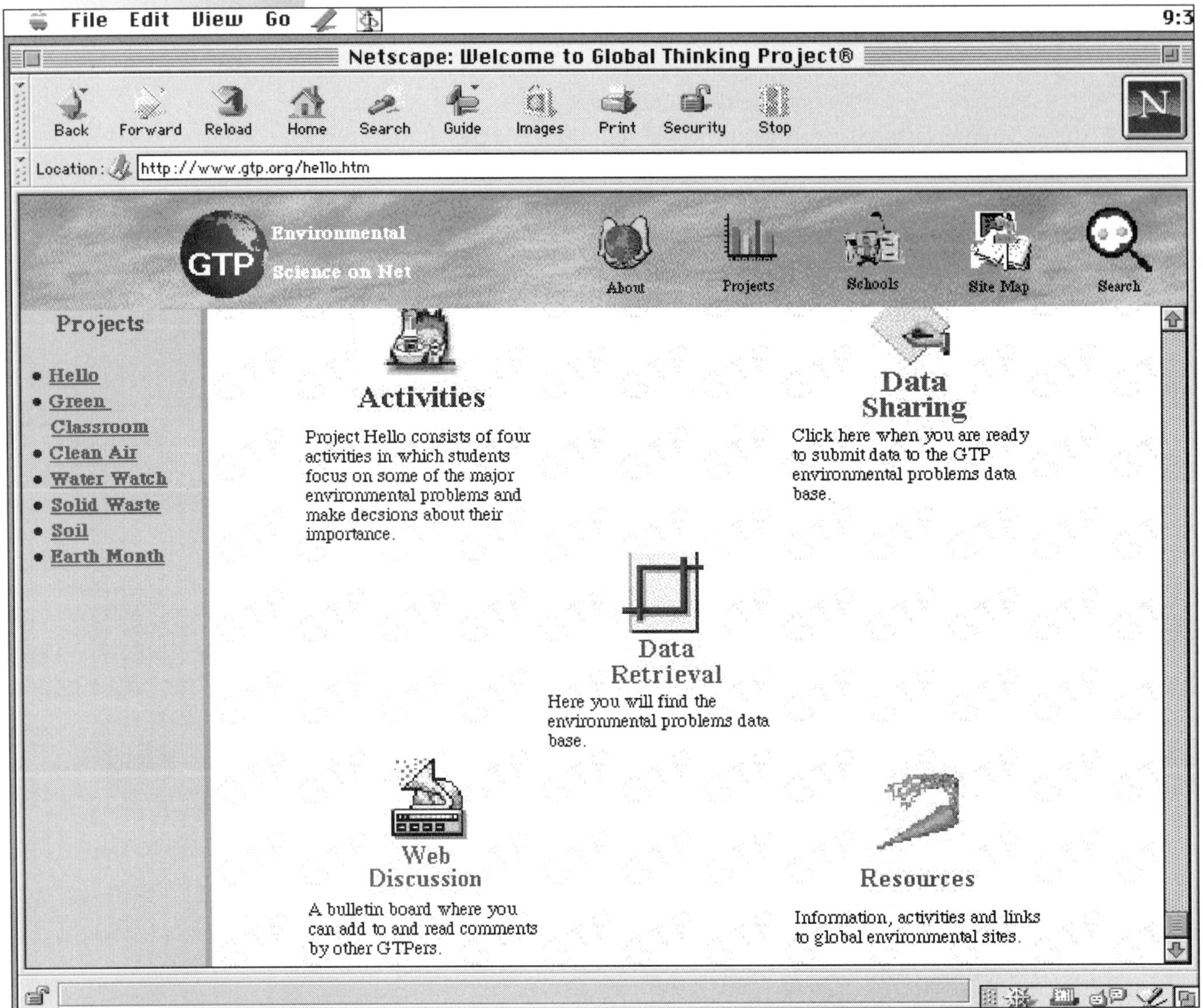

Project Hello Home Page

Data Retrieval and Analysis

1. From the Project Hello home page, select "Data Retrieval." Allow each team time to obtain comparable data from other schools completing this project. Record the data from other schools on a large class chart.
2. After teams have had a few minutes to look over the data, instruct them to write one or two questions they could answer using the data available.

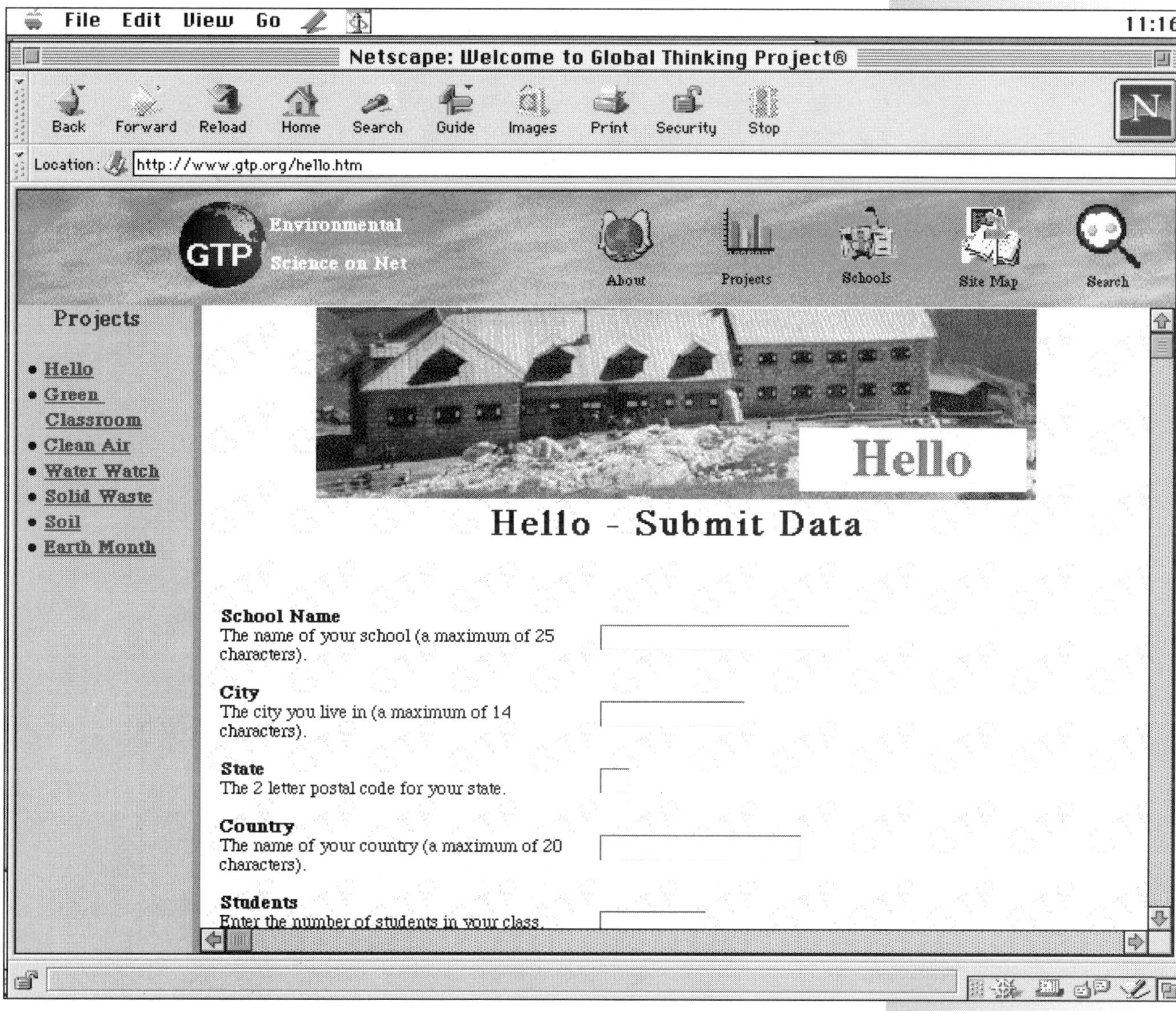

Project Hello Data Entry Form

3. Help each team select one question, and prepare a report for the class in which they state their question, present relevant data, and draw conclusions based on the data. Give teams time to use the computer to make graphs and complete appropriate analyses. This report should be prepared in the form of a poster to be presented to the rest of the class.

Web Discussion

From the Project Hello home page, select "Web Discussion." Post your findings, comments, and questions. Use this forum to obtain further information for your research, if desired. Respond to other teams' comments and questions.

School	City	State	Country	Number of Students	Rank
gsu	atlanta	GA	U.S.A.		1
GSU	Atlanta	GA	U.S.A.	20,000	12
gsu	atlanta	ga	usa	11	6
gsu	athanta	ga	usa	111	1
Municipal School 1	Yaroslavl		Russia	20	4
Lafayette Middle	Lafayette	GA	USA	28	3
Lyecee 93	Chelabinsk		Russia	35	6
School 157	St. Petersburg		Russia	26	4
Escola Lavinia	Barcelona		Spain	19	8
Salem High School	Conyers	GA	USA	26	4
Melbourne Girls Grammar School	Melbourne		Australia	29	7

Project Hello Data Retrieval Form

Optional Extensions for Project Hello

Here are additional activities you might want to use with your class:

1. Create a GTP resource library for your classroom. Each team should find different articles, paste them on construction paper, and write one- or two-sentence summaries or questions about its articles. Materials managers should retrieve the newspapers and magazines, construction paper, scissors, and glue for their team. Checkers are responsible for facilitating work on the task. When completed, the tracker should post the articles on the Global Thinking Bulletin Board in the classroom, and organize the reports in file folders in a class reference library.

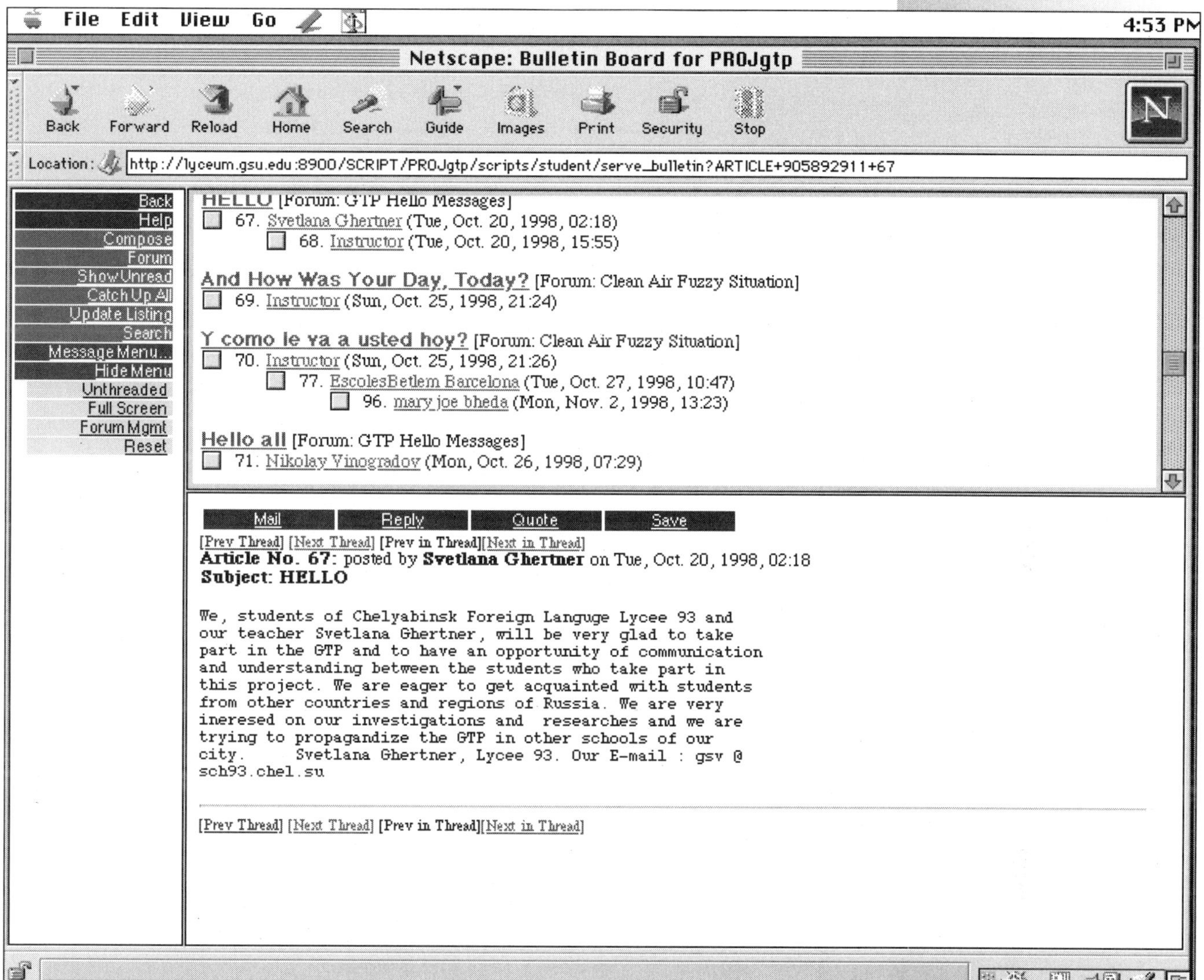

Sample "Discussion" Message

2. Select a topic of interest, such as family life, school life, sports, teen culture, religion, food, government, customs and beliefs, or holidays to profile in depth and share with other schools participating in your global community. Design a survey instrument to collect information from all the members of the GTP, and share your findings in the GTP listserve or the GTP bulletin board.
3. Construct a climatogram, a graph that shows the average monthly rainfall vs. temperature, for your city or town. Over what period of time were these data collected? Are any trends evident? Has the climate changed in any significant way over time? Speculate about why. Have the climatic changes had any effects on plant and animal life in your region?

4. Read newspapers, listen to the radio, watch television, and interview interested citizens to learn about some of the major ecological problems in your town or region. Discuss your findings with other members of your global community and post your results on the GTP listserve so others can discuss your findings with you.
5. Study the attached quotations from astronauts who have viewed Earth from space. What elements of these astronauts' thinking are found in your class list of global thinking statements? Are there any other new ideas? Add them to your list.

   "For those who have seen the Earth from space, and for the hundreds and perhaps thousands more who will, the experience most certainly changes your perspective. The things that we share in our world are far more valuable than those which divide us." (Donald Williams, USA)

   "After an orange cloud—formed as a result of a dust storm over the Sahara and caught up by air currents—reached the Philippines and settled there with rain, I understood that we are all sailing in the same boat." (Vladimir Kovalyonok, USSR)

   ". . . During the eight days I spent in space, I realized that mankind needs height primarily to better know our long-suffering Earth, to see what cannot be seen close up. Not just to love her beauty, but also to ensure that we do not bring even the slightest harm to the natural world." (Pham Tuan, Vietnam)

   "From space I saw Earth—indescribably beautiful with the scars of national boundaries gone." (Muhammad Ahmad Faris, Syria)

   "Before I flew I was already aware of how small and vulnerable our planet is; but only when I saw it from space, in all its ineffable beauty and fragility, did I realize that humankind's most urgent task is to cherish and preserve it for future generations." (Sigmund Jahn, GDR)

   "The first day or so we all pointed to our countries. The third or fourth day we were pointing to our continents. By the fifth day we were aware of only one Earth." (Sultan Bin Salman al-Saud, The Kingdom of Saudi Arabia)

   ". . . We are one world." (John-David Bartoe, USA)

6. As a class, develop a questionnaire to assess the extent of global thinking in younger or older students or adults in the community. Develop a strategy for collecting and analyzing the data obtained from the survey and for sharing the results with other schools in the project. Invite other schools in the GTP to participate in your study. Post the results on the GTP listserve or GTP bulletin board.
7. Assign each team a continent or bioregion. Teams should select a country in their continent or bioregion, and research its imports, exports, and trading partners. On a map of the world, connect each country with its trading partners (e.g., red string for exports, blue string for imports). Discuss as a class how economic interdependence is related to the global thinking concept.

CHAPTER 3

Project Green Classroom

. . . all education is environmental education, by which I mean the students are taught in various and often unintended ways that they are part of or apart from natural systems.
David W. Orr
Environmental Educator

How would you rate the environmental quality in your school? your classroom? What kinds of observations can you make to help you answer this question? How do the conditions in your classroom compare with those in other classrooms around the world? In this activity, students investigate some of the factors contributing to the environmental quality of their classroom, and compare their results with those obtained in other Global Thinking classrooms.

In Project Green Classroom, students work together to improve the environmental quality of their classroom.

Unlike the other projects in this book, Project Green Classroom is one large activity, broken down into five independent investigations to be performed by different teams of students. Divide the class into teams of four or five students, and remind them to assign and rotate roles (See page xiv for role cards.) during the project. The entire project can be completed in about five class periods.

Planning Alert
Set out particulate collectors seven to ten days before beginning this project. (See Resources for suppliers.)

Expose agar plates three to four days before beginning this project. (See Resources for suppliers.)

Instructions for advance preparation can be found on pages 45 and 46.

Goals for Project Green Classroom

1. Students will monitor and evaluate the environment in their classroom.
2. Students will compare and contrast their observations with those in other GTP schools.

Planning Chart

Team Activity	Class Periods
Gather "Green" data.	1–2
Compile and discuss class data.	1
Collect and discuss "Green" data from other schools; prepare poster reports.	2–3
Present poster reports.	1

Web Site Information

From GTP home page (www.gtp.org) click on "projects" to access Project Green Classroom home page.

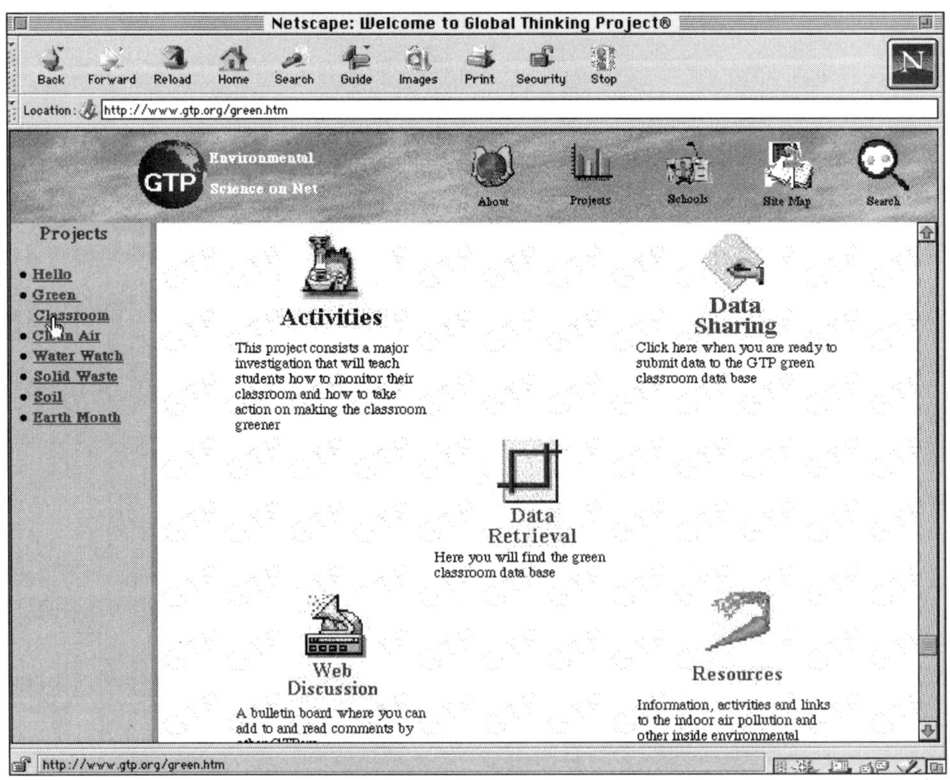

Project Green Classroom Home Page

ACTIVITY 3.1

How Green Is Our Classroom?

Objectives

- Explain what is meant by *environment.*
- Examine the quality of the classroom environment.
- Compare and contrast the environmental quality of the classroom with the environmental quality of classrooms in other communities.
- Develop and implement strategies for improving the classroom climate.

Materials

thermometer, sling psychrometer, light meter, measuring tapes, hand lenses, mm-ruled graph paper, bathroom scale, paper sacks, microbiology reference books, particulate collectors*, nutrient agar plates**, team assignments from pages 47–49, Student Log pages 196–201.

*Advance Preparation of Particulate Collectors: About seven to ten days before the activity, prepare particulate collectors by cutting two windows in index cards. Cover each window with clear sticky tape. Hang or tape three or four cards in different locations around the room (e.g., near the window or door, in a quiet corner, on the teacher's desk). Expose the collectors to the air for at least seven days. Cut pieces of 1 mm^2 graph paper and place them over the tape on the sticky side. Use a microscope to view the particles collected. The 1 mm^2 graph paper can be used as a scale while viewing the particulates.

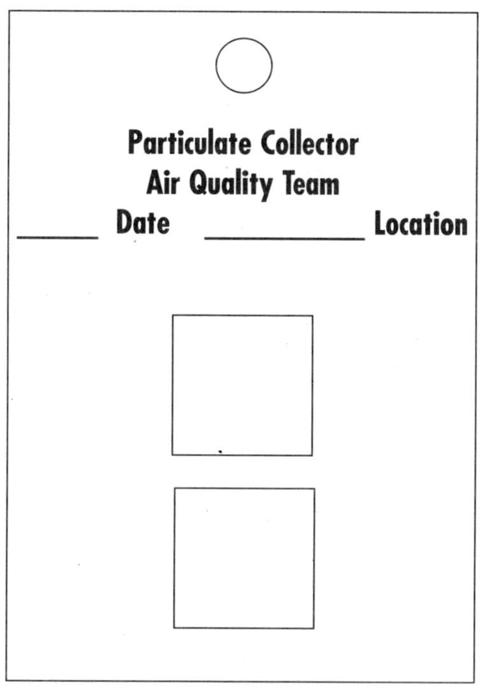

Particulate Collector Template
Use this design as a template and make copies for your students.

**Advance Preparation of Agar Plates: Obtain or prepare four or five sterile petri dishes containing nutrient agar, and number them on the bottom. Leave dish #1 covered (the control). Open the other dishes and expose them to the air for fifteen or twenty minutes in different areas of the classroom (e.g., near the window or door, in a quiet corner, on the teacher's desk). Then cover the dishes and tape them closed with transparent tape. Record the date, location, and time exposed for each dish, and incubate the plates upside-down at room temperature for three to four days.

Procedures

1. Begin the lesson by asking the students how they would rate the level of environmental quality in their classroom. What criteria are they using to make their evaluations? Brainstorm as a class a list of factors that might contribute to or reflect the level of environmental quality in your classroom. Write these suggestions on a piece of chart paper and display them in the classroom.
2. Explain to the students that they are going to conduct a survey to examine the environmental quality of their classroom. Each team will have a different task and will contribute to the final class report. Give each team a task (See pages 196–201 in the Student Log.), and allow time for the students to complete the surveys. Remind team members to use their roles, so everyone has a chance to participate.

Team Assignments

- Weather Team
- Air Quality Team
- Physical Plant Team
- Population Team
- Microorganism Team
- Trash Team

3. After each team has compiled its results, the communicator should report the data according to the report format shown at the bottom of their team's data sheet, and enter it into a class data chart. When all of the teams have reported, return to the question that opened this activity. How would the students evaluate the environmental quality of their classroom now? What evidence can they bring to bear that was not available before? What new questions do they have? What ideas do they have for improving the environmental quality of their classroom? Have the communicators from each group meet together to summarize the class discussion in a classroom environmental profile. Publish your class profile on the Project Green Classroom home page under Web discussion.

Weather Team Assignment

Weather Team

You will need: thermometer, sling psychrometer, camera with a light meter

Your task is to make simple measurements that will allow you to make statements about the classroom's climate over the course of a day.

1. Measure the air temperature of the room. Do you think it is the same everywhere in the room? How could you find out? Does distance from the floor matter? Devise a way of measuring the air temperature at a constant height at different points in the room.

   Is the air temperature in the classroom constant during the day? Measure the air temperature at the same place in the room at intervals during the day.
2. Use the sling psychrometer to measure the relative humidity in the classroom. If you can, make these measurements throughout the day, at the same intervals at which you measure the air temperature.
3. Use the light meter to measure the light intensity in the room. If the classroom has many windows, you might want to make several measurements at different distances from them.
4. Measure the air temperature, relative humidity, and light intensity just outside your classroom. Compare the indoor and outdoor readings.

Air Quality Team Assignment

Air Quality Team

You will need: particulate collectors (prepared in advance by your teacher), microscope or hand lens, mm-ruled graph paper

Your task is to examine particulates that have been collected from the air in your classroom over a period of a week.

1. Examine the slides that have been placed about the room during the past week to collect particulates. Using a microscope or hand magnifier, determine:
   - The average number of particles per square mm
   - The types of particles (small, medium, large; colors; shapes)
2. Are there differences depending on where the collectors were located? What hypotheses can you make to account for differences? If you wish, put out several new collectors to test your hypotheses.

Physical Plant Team Assignment

Physical Plant Team

You will need: paper and pencils, measuring tapes

Your task is to map the physical parameters of the classroom.

1. Measure the length, width, and height of the classroom. Calculate its volume.
2. Draw a scale floor plan of the classroom, indicating locations of windows, doors, heating/air conditioning vents, furniture, plants, and anything else visible in the room. How many square feet of wall are devoted to windows? (What fraction of the total wall space is this?)
3. What type(s) of light fixtures are used? How many hours/day are they used? How many kilowatt hours of electricity are consumed lighting your classroom? How much does this cost the school system?

Population Team Assignment

Population Team

You do not need any special equipment.

Your task is to determine how the classroom is used by macroscopic living things.

1. How many human beings occupy this room during a day? How long do they stay? Count the number of people who go in and out of the classroom during the day, and how long each stayed. Calculate the number of person-minutes the classroom was occupied.
2. How many plants are there in your classroom? of what types? Are they large or small? Do they appear to be healthy? (What criteria are you using to assess their health?)
3. How many animals are there in your classroom? of what types? Are they large or small? Do they appear to be healthy? (What criteria are you using to assess their health?)

Microorganism Team Asssignment

Microorganism Team

You will need: agar plates exposed in different parts of your classroom (prepared in advance by your teacher), dissecting microscope or hand lens, reference books

Your task is to examine the types of microorganisms that can be collected in different parts of your classroom.

1. Without opening them, examine the microorganisms growing in the petri dishes. Sketch and describe the different types of colonies you observe.
2. Use reference books to determine whether your colonies are mold or bacteria (mold colonies are generally fuzzier and larger than bacterial colonies).
3. Were different microorganisms present in different areas of the classroom? Develop some hypotheses for your results.

Trash Team Assignment

Trash Team

You will need: one day's trash collected from your classroom, paper sacks, digital bathroom scale

Your task is to investigate the trash produced in your classroom during a typical day.

1. Sort the trash collected for one entire day in your classroom into categories: paper and paperboard, metals, glass, plastics, rubber and leather, textiles, food wastes, other.
2. Place each collection in a paper sack and weigh it on the bathroom scale. (One way to do this is to weigh one team member with and without each sack.)
3. Calculate the total mass of trash collected in your classroom in one day and the percent of each category of trash present.

Classroom Green Data Chart Sample

Category	Information	Your Class	School 1	School 2	School 3
School	name				
Town	name				
Date	day-month-year				
Latitude	hh:mm n/s				
Longitude	hh:mm e/w				
Temp. inside	average temperatures				
Temp. outside	average temperatures				
Relative humidity	%				
Light intensity					
Room length	meters				
Room height	meters				
Room width	meters				
Room volume	cubic meters				
Population	number of people in room/day				
Plants	total number in room				
Animals	total number				
Particles	average number of particles/mm				
Particle size	average particle size in room				
Microorganisms	yes, if some observed; no, if not.				
Paper	% of total				
Metal	% of total				
Glass	% of total				
Plastic	% of total				
Rubber/leather	% of total				
Textiles	% of total				
Food	% of total				
Other	% of total				
Trash	total mass in grams				

Using the World Wide Web

Data Transmission

After the data have been compiled, the class should submit its data to the GTP Green Classroom database. From the Project Green Classroom home page, select "Data Sharing." Students will be able to enter their data directly into a Web form and send it electronically to the GTP database.

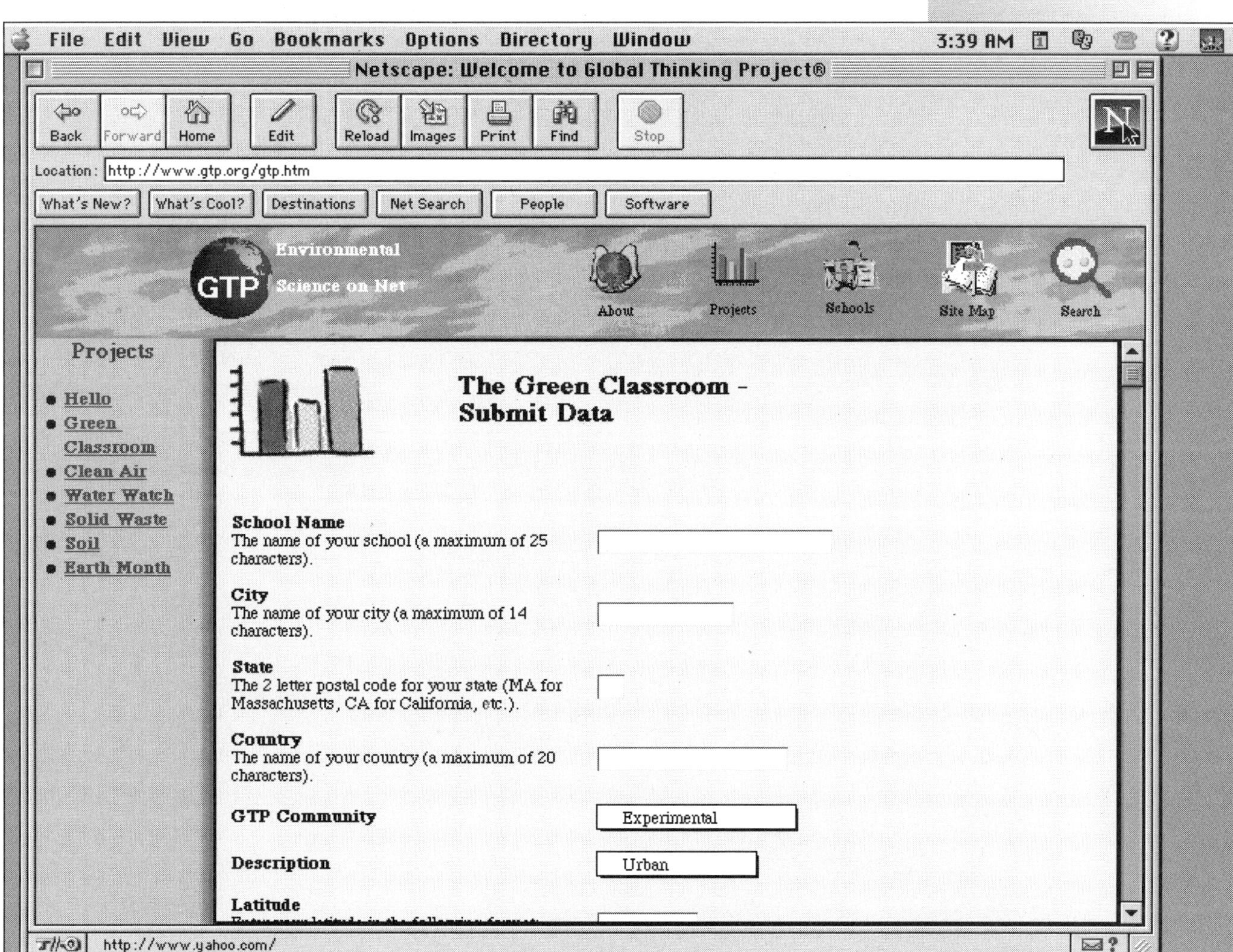

Web Data Form

Data Retrieval and Analysis

1. From the Project Green Classroom home page, select "Data Retrieval." Allow each team time to obtain comparable data from other schools completing this project. Have each team record the data from other schools on the class's Green Classroom Data Chart (Student Log page 202).

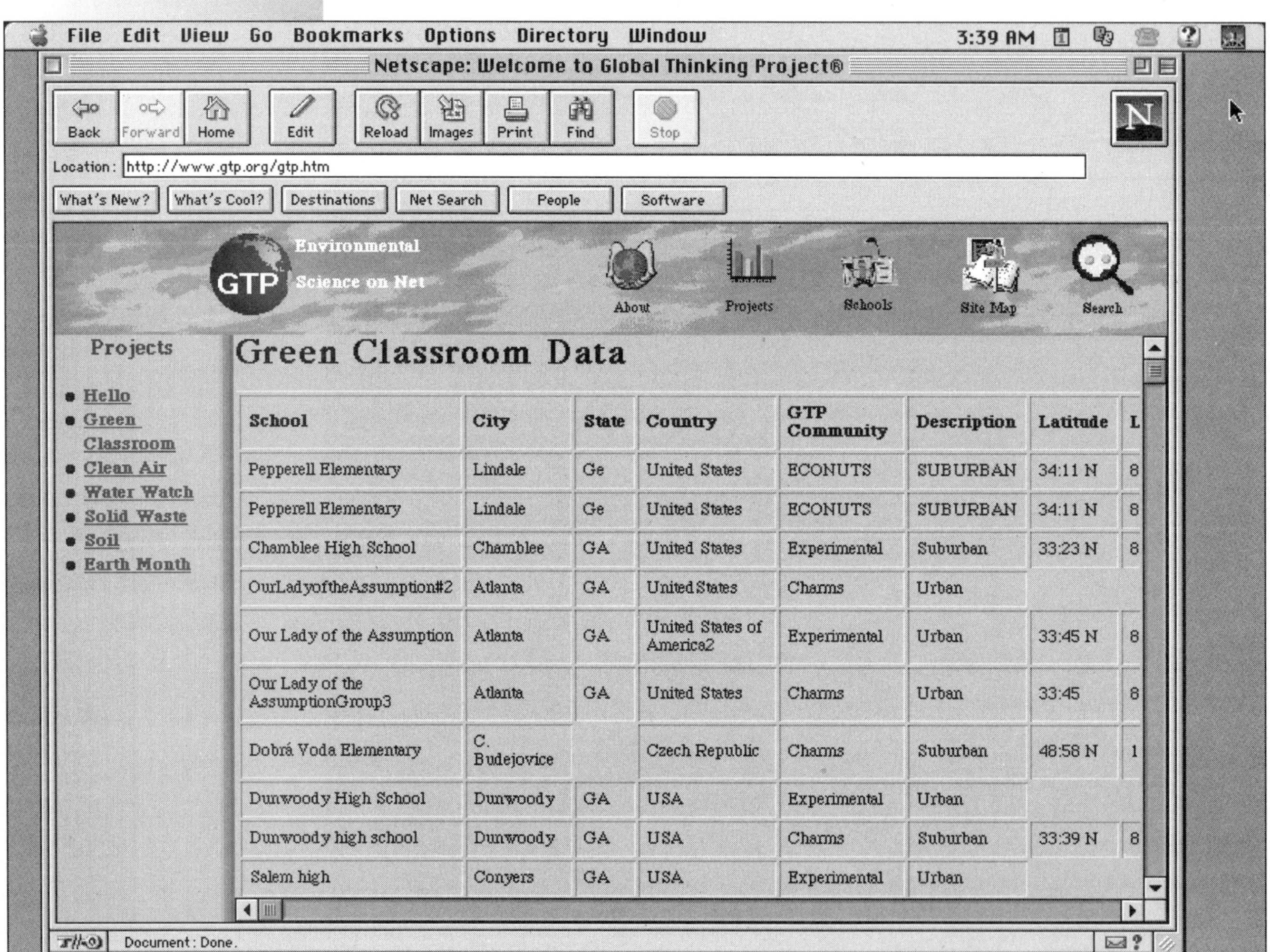

File Edit View Go Bookmarks Options Directory Window 3:39 AM

Netscape: Welcome to Global Thinking Project®

Back Forward Home Edit Reload Images Print Find Stop

Location: http://www.gtp.org/gtp.htm

What's New? What's Cool? Destinations Net Search People Software

GTP Environmental Science on Net

About Projects Schools Site Map Search

Projects

- Hello
- Green Classroom
- Clean Air
- Water Watch
- Solid Waste
- Soil
- Earth Month

Green Classroom Data

School	City	State	Country	GTP Community	Description	Latitude	L
Pepperell Elementary	Lindale	Ge	United States	ECONUTS	SUBURBAN	34:11 N	8
Pepperell Elementary	Lindale	Ge	United States	ECONUTS	SUBURBAN	34:11 N	8
Chamblee High School	Chamblee	GA	United States	Experimental	Suburban	33:23 N	8
OurLadyoftheAssumption#2	Atlanta	GA	United States	Charms	Urban		
Our Lady of the Assumption	Atlanta	GA	United States of America2	Experimental	Urban	33:45 N	8
Our Lady of the AssumptionGroup3	Atlanta	GA	United States	Charms	Urban	33:45	8
Dobrá Voda Elementary	C. Budejovice		Czech Republic	Charms	Suburban	48:58 N	1
Dunwoody High School	Dunwoody	GA	USA	Experimental	Urban		
Dunwoody high school	Dunwoody	GA	USA	Charms	Suburban	33:39 N	8
Salem high	Conyers	GA	USA	Experimental	Urban		

Document: Done.

Project Green Classroom Data Table

2. After teams have had a few minutes to look over the data, instruct them to write down at least five questions they could answer using the data they have available.
3. Help each team select one question and prepare a report for the class in which they state their question, present relevant data, and draw conclusions based on the data. Give teams time to use the computer to make graphs and complete appropriate analyses. This report should be prepared in the form of a poster to be presented to the rest of the class.

Model Middle School's Green Report
Global Thinking Project

1:52 P.M. November 9, 1995
From: model@igc.apc.org (Model Middle School)
Date: Thu, 09 Nov 95 17:47:59 EST
Hello Econuts:

We have finished our study on "How green is our classroom?":

Here is our data:

Model Middle School, Rome, Georgia, 9-18-95, N 34 degrees 18'30", W 85 degrees 06, 21 celsius, 23 celsius, 69%, X, 13.55m, 2.43m, 3.96m, 130.39, 41, 0, 6, less than 1 (24 per cmxcm, 1mm, yes, 97%, 2%, 0%, less than 1%, less than 1%, 0%, 0%, less than 1%, 733g.

Here are brief comments by each of the teams in the project:

Weather Team: We measured our weather between 12:45 and 1:00 inside and outside. Our data is about one month old now. It is November now and much cooler.

Air Quality Team: Our whole class participated in this measurement. We placed 4 particulate collectors around the room. After one week, all had particulates. Some particulates were long fibers of different colors. When we looked with a scope with a greater magnification, we could see many particles that were very numerous. It was difficult to give numbers of large, medium, and small particles since there were no measurements to go by. We made up our own: large is greater than 3mm, medium is 13mm, and small is less than 1mm. We found that the number of particles per square cm is about 70. That is equal to 0.7 per mm squared. The average particle size was less than 1mm and we could see several with a microscope.

The physical plant discovered with meter sticks that our room is 13.55m long, 2.43m high, 3.96m wide, and the volume of the room is 130.39 cubic meters, and the lighting we have 10 fluorescent lights that burn 7 hours a day and we do not know kwh of electricity consumed.

The Population Team: We discovered a time interval of 330 minutes. In that time a total of 41 people came into the room. We have an aquarium with five fish in it. They are brim from a pond in our area. They have grown a lot since we put them in the tank.

Microorganism Team: We discovered that in every location, bacteria was present in the air. We found about 6 colonies of mold in each of our different locations. There was an average of 57 bacteria colonies in each dish around the room.

Physical Plant Team: We have just gotten a new mobile unit for our classroom. We were unable to measure light intensity.

Trash Team: We had 97 percent paper in our trash. We had no food, glass or textiles. We had so much paper trash because we just cleaned out our aquarium to set it up for the year.

Student Green Report Sample

Web Discussion

From the Project Green Classroom home page, select "Web Discussion." Post your Classroom Environmental Profile or the student teams' research projects for discussion and comment. Respond to other postings with your team's comments and questions.

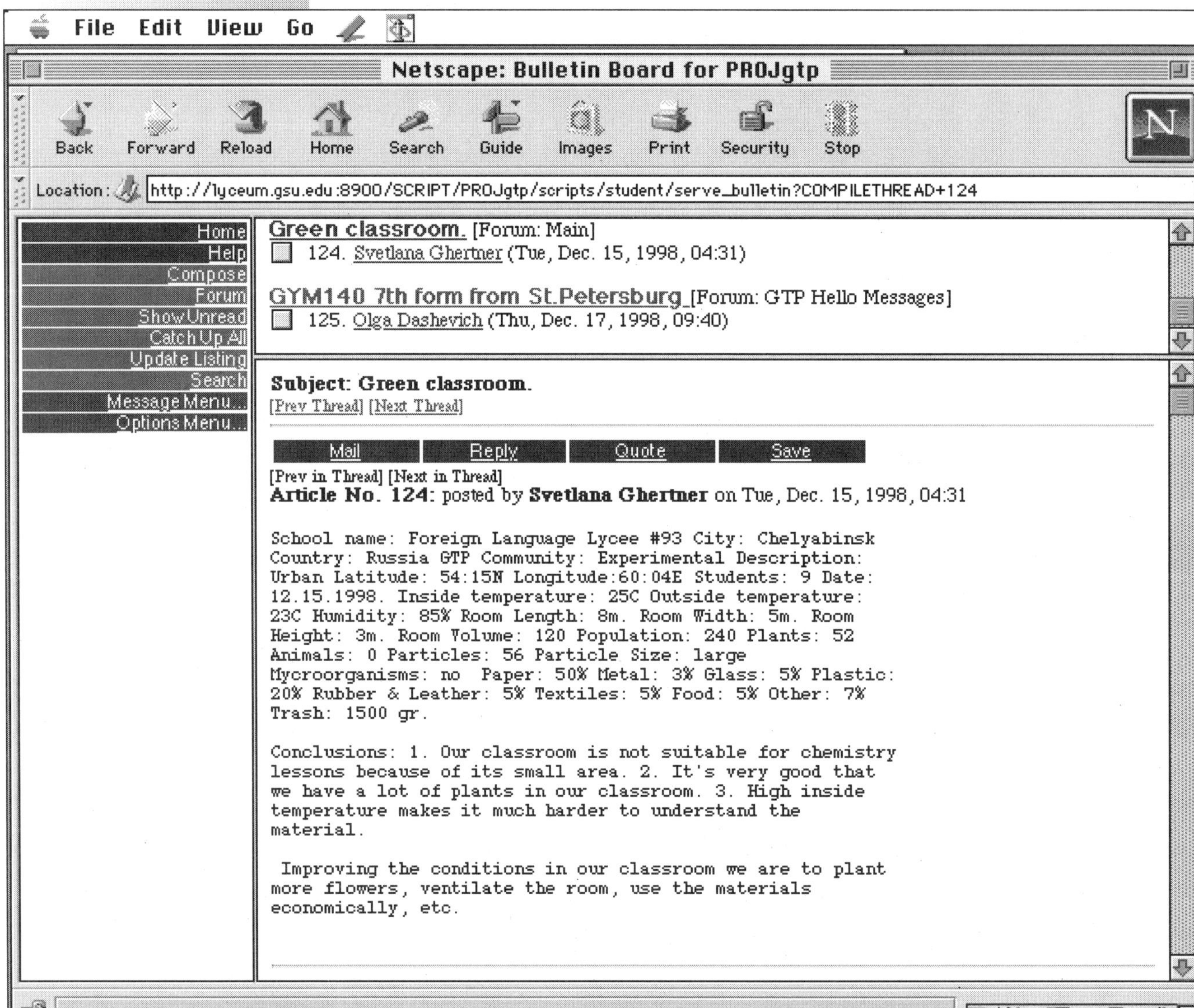

Sample "Discussion" Message

Optional Extensions for Project Green Classroom

Here are additional activities you may want to use with your class.

1. Develop and implement an action plan for improving your classroom's environmental quality. Assess its quality six months from now and note any improvements or deteriorations.
2. Survey the school or your home for evidence of environmental quality. Suggest and implement strategies for improving the environmental quality of these locations.

CHAPTER 4

Project Clean Air

The voices of my grandfather said to me, "The air is precious. It shares its spirit with all the life it supports. The wind that gave me my first breath also received my last sigh. You must keep the land and air apart and sacred, as a place where one can go to taste the wind that is sweetened by the meadow flowers."

Chief Seattle, Northwest Nations

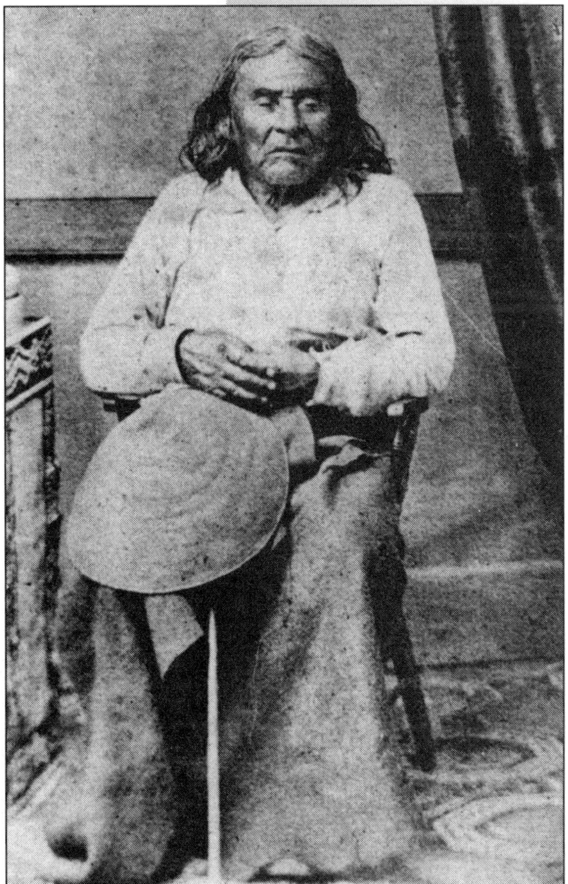

Chief Seattle

The atmosphere is a very thin layer of gases and solid particles that surrounds the Earth. Since the atmosphere knows no national boundaries, the air in one region can be greatly influenced by actions in a distant area. Recently the U.S. Environmental Protection Agency concluded that current air quality standards should be strengthened. In this project your students will investigate air quality, learn how to monitor the air, and learn how air quality is measured.

This project will serve as an example of the principle that global problems have local origins and that careful monitoring and study of the local environment is the first step in being a global thinker.

In this investigation of the air around them, students will look at a primary and a secondary pollutant. *Primary pollutants* enter the atmosphere from identifiable sources, such as pollen from plants, dust from soil, or carbon monoxide from vehicle emissions. *Secondary pollutants* form from chemically altered primary pollutants, as in the case of ozone. There are natural sources of both primary and

Primary pollutant is any substance released into the atmosphere in harmful concentrations.

Secondary pollutant is a pollutant resulting from the chemical reaction between two or more primary pollutants, or a primary pollutant and a naturally occurring atmospheric element.

secondary pollutants. Volcanoes, forest fires, plants (pollen and the giving off of gasses), windblown dust, and breaking waves are examples of natural sources of primary pollutants.

Major Air Pollutants

Pollutant	Description	Sources	Effects
Carbon Monoxide (CO)	colorless, ordorless gas	• vehicle burning gasoline • kerosene or wood-burning stoves	• headaches, reduced mental alertness, death • heart damage
Lead (Pb)	metallic element	• vehicles burning leaded gasoline • metal refineries	• brain and kidney damage • contaminated crops and livestock
Nitrogen oxides (NOx)	several gaseous compounds made up of nitrogen and oxygen	• vehicles • power plants burning fossil fuels • coal-burning stoves	• lung damage • acid rain • forests damaged • ozone and other pollutants (smog) formed
Ozone (O_3)	gaseous pollutant	• vehicle exhaust and certain other fumes • formed from other air pollutants in the presence of sunlight	• lung damage • eye irritation • respiratory tract problems • vegetation damaged • smog
Particulate matter	very small particles of soot, dust, or other matter, including tiny droplets of liquids	• diesel engines • power plants • industries • windblown dust • wood stoves	• lung damage • eye irritation • crops damaged • visibility reduced • buildings and statues discolored
Sulfur dioxide (SO_2)	gaseous compound made up of sulfur and oxygen	• coal-burning power plants and industries • coal-burning stoves • refineries	• eye irritation • lung damage • aquatic life killed • acid rain • forests damaged • buildings and statues deteriorated

Particulates are solid particles or liquids in the atmosphere. Particulates can settle on surfaces. Particles can provide surfaces on which water condenses. This can increase fog, clouds, and rain depending upon weather conditions. Particulates can also increase acid rain when in the presence of sulfur oxides; this happens because the particulates act as cloud condensation nuclei, which then form droplets. The sulfur oxides then readily dissolve into existing droplets. Particulates can substantially reduce visibility, which can increase accidents. Particulates will reflect incoming solar radiation.

Ozone is a secondary pollutant formed by a chemical reaction between gases in the presence of high temperatures and sunlight. To monitor these air pollutants, students will use simple measuring devices. In the case of particulates, they will capture solid particles using an index card with a hole covered with sticky tape. Ozone will be measured using a chemically sensitive paper strip—the Eco-Badge. (See Resources for suppliers.)

In Activities 4.2 to 4.4, students learn how to monitor the atmosphere. (See the planning chart below to see how time for the activity should be allotted.) In the last activity, Activity 4.5, they use this knowledge to design an air pollution project.

Goals for Project Clean Air

1. Students will investigate types of air pollution.
2. Students will learn techniques to monitor particulates and ozone found in the lower atmosphere.
3. Students will design a research project based on questions and inquiries about air quality.

Planning Chart

Activity	Class Periods
1. A Breath of Outdoor Air	1
2. Monitoring the Air	1
3. Monitoring Particulates	1
4. Monitoring Ozone	2
4. Analyzing Ozone Data	1
5. Air Pollution: What Can You Do About It?	2–3

Planning Alert
You will have to prepare particulate collectors (see page 45) in preparation for Activity 4.3. The counters will need to be placed in locations for at least seven days. You should begin Activity 4.4 as soon as the particulate counters are set up. In Activity 4.4, students will do a mini-research study at home where they measure ozone levels and other atmospheric variables.

Web Site Information

From the home page (www.gtp.org), click on "projects" to access Project Clean Air home page.

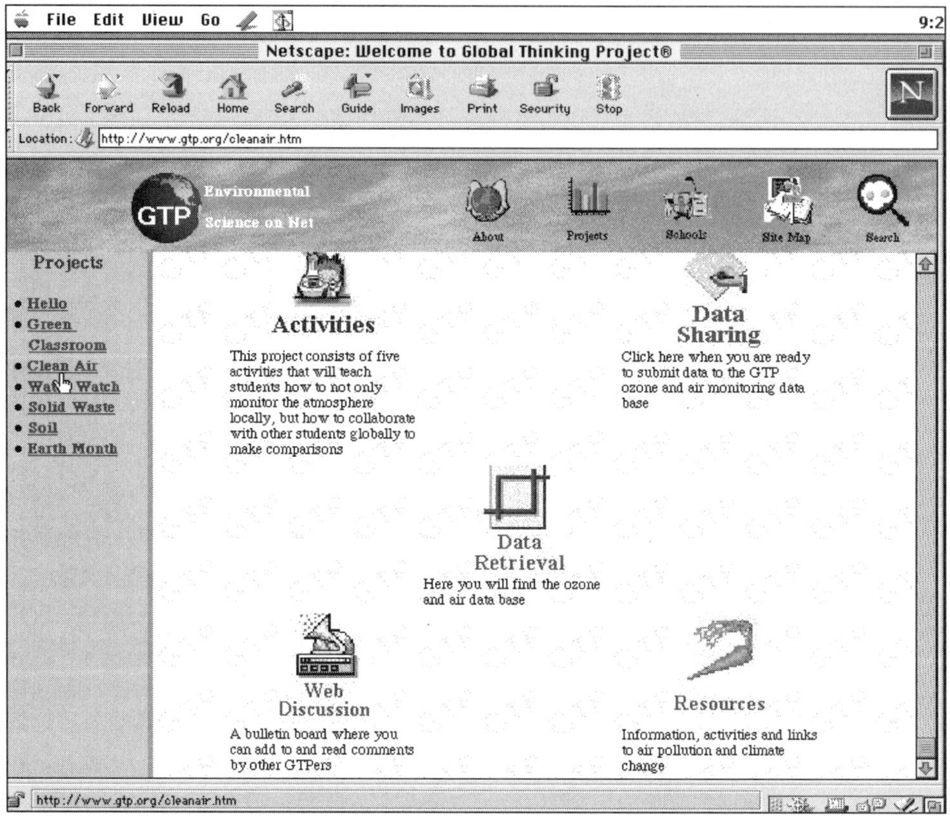

Project Clean Air Home Page

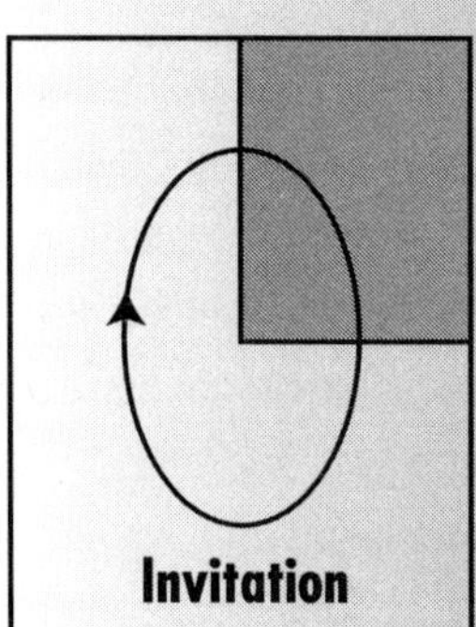

ACTIVITY 4.1

A Breath of Clean Air

Students are invited to share what they know about air pollution by participating in a brainstorming activity. They work in small teams to discuss and create a chart of what they have heard and questions they have about air pollution. The results are posted on charts and hung in the classroom for referral throughout this project.

Objectives

- Assess students' prior knowledge of air pollution.
- Generate questions about air pollution.
- Identify causes of atmospheric change.
- Identify reasons why atmospheric monitoring is important to our understanding of weather and climate change.

Materials

chart or construction paper, crayons or marking pens, hand lenses, masking tape, pieces of sticky tape that have been placed outside for a week, an article about air pollution from a newspaper or magazine (such as a story about a smog alert, volcanic eruption, effects of acid rain, pollen and health, ozone); Student Log pages 205–207

Advance Preparation

One week prior to the first lesson, you should collect examples of atmospheric suspended *particulate matter.* You can do this very easily by cutting a small square in index cards and then placing sticky tape over the square. Place about six of these outside in an unobstructed area for two days. Solid particles—dust, pollen, soot—will be deposited on the sticky tape.

Procedures

1. Divide your class into teams of four students. You should use a random method when forming the teams, and inform the students that they will be working together for the duration of Project Clean Air. Give each team an index card that has a piece of sticky tape that has been placed outside for a week. Tell the students to observe the card with the hand lens and describe what they see. What do the specs look like? Can they identify them? How many are on the card? What do they think is the source of these objects? Give the teams five minutes to make their observations and answer these questions. Follow up by asking students from some of the groups to report their team's findings.

2. Tell the students that they are going to begin a study of air pollution, from a local as well as a global perspective. Read a few paragraphs from an article on air pollution (find one from a local newspaper or magazine) to highlight the nature of air pollution and problems associated with polluted air. Discuss the article with the students. If you can't find an article, here is an excerpt from an article entitled "Gasping for Clean Air."[1]

[1] Wasserman, P., and Doyle, A. *Earth Matters.* Washington, DC: Zero Population Growth, pp. 41–43, 1992.

Air pollution
is any toxic or radioactive gases and particulate matter introduced into the atmosphere, principally as a result of human activity. Pollutants can be carried by winds and rising air currents, with the larger particles returning to Earth's surface under gravity, while smaller particles are removed by rain.

Particulate matter
is solid particles and liquid droplets present in the atmosphere. Particulate matter has various sources, both natural and as a result of human activity. Examples include soot and ash emitted from factories, dust raised by agricultural and mining activities, and dust and ash generated by volcanic eruptions.

"In many cities, it is actually hazardous to breathe. In the United States, some 150 million people breathe air considered unhealthy by the U.S. Environmental Protection Agency (EPA). Breathing air in Bombay, India, is equivalent to smoking ten cigarettes a day. And in Mexico City, the world's most polluted and populated city, infectious diseases like salmonella and hepatitis can be contracted simply by inhaling bacteria suspended in the air."

3. To find out what your students have heard about air pollution and what questions they have, conduct this very brief activity with them. Show students a T-chart like the sample on page 61 drawn on a large sheet of newsprint paper, and ask them to brainstorm answers to two questions:

   - What have you heard about air pollution?
   - What questions do you have about air pollution?

   Give students ten minutes to generate a list of answers to these questions. When the teams are finished, tell them to look their lists over again and select the most important item on each list. Have them circle these items on their chart paper. The communicator should record their team's decisions on a master chart on the chalkboard or on chartpaper. Discuss the results with the class and use the discussion to focus on air pollution, its causes, and what can be done to maintain "clean air." Hang the individual team T-charts in the room. You will refer to these throughout the project.

   Note: The purpose of this activity is to enable students to talk about their initial ideas of air pollution. It is not necessary to "correct" these initial ideas, but to make use of their ideas as you continue with this project. The questions that the students list can be used in Activity 4.5 in which they design a project.

Air Pollution T-chart Sample

What have you heard about air pollution?	What questions do you have about air pollution?

4. Ask the students to think and write about the following questions:
   - How does the quality of the air impact you directly?
   - How does it affect other living things in the environment?
   - What is "clean air," and what causes the air's quality to change?
   - How do you think the quality of the air in your region compares to regions close to you, as well as in places quite far from you?

   Give each team a card with a different question on it. Tell the team to discuss the question for two minutes and write their ideas on a separate sheet of paper. Exchange cards so that each team gets a new question. Continue the process so that each team has an opportunity to discuss each question. Discuss the students' ideas with the whole class. To maintain a sense of positive interdependence, randomly call on students from the teams to respond to the questions. Make sure everyone in each team has a number (1, 2, 3, and so on). Draw a number from a set of index cards that have the numbers 1–4. The person in each group whose number is drawn stands and responds to the question for the group. Repeat the process as you discuss each question. At the end of the process have students write in their journals about each question.

Exploration

ACTIVITY 4.2

Monitoring the Air

In this activity students will be introduced to the concept of monitoring and learn how to monitor important atmospheric variables such as temperature, humidity, cloud cover, cloud type, visibility, wind, and the weather. Students summarize their results and send weather reports to schools in the GTP.

Objectives

- Explain why monitoring the quality of the air is important to students as well as other living things and their environment.
- Collect and organize data on several atmospheric variables.
- Send weather reports to other schools.

Materials

thermometers, dowel, screw, small pieces of cloth, rubber bands, water, relative humidity charts from a science textbook, Dwyer wind meter

Advance Preparation

You'll need to take your class outside briefly during this lesson to monitor the air. Select a location in advance that is conducive to viewing the sky and making temperature and humidity measurements. Students will have to revisit this location for several days in order to compile a week of observations.

Procedures

1. Have your students imagine that they are members of air quality research teams and that their job is to monitor the air in their local environment and be prepared to share their research results with other student research teams. Tell them they will use the Internet to share and collaborate with students in the GTP.
2. Ask the class, "What are some factors about the air that might be important for an air quality research team to monitor?" Give each team three to five minutes to jot down their ideas. Write the teams' suggestions on a master list on the board. Using the ideas the students have suggested, work with the class to come up with a set of observations that you and they think are important.
3. Help the class generate a master list of observations. Create a class data table that students can use to record their data. Here are some observations that you should include:

Weekly Air Monitoring Table

Row	Air Temp.	Rel. Hum.	Cloud Cover	Sun-shine	Wind Speed	Wind Direc.	Precip.	Lat.	Long.	School	Date
1											
2											
3											
4											
5											

4. Discuss the types of observations and measurements that should be made for each of the variables.
   Air Temperature: Use a thermometer and measure in degrees Celsius.
   Humidity: Report the relative humidity. Use a psychrometer, which is simply two thermometers, one of which is a "wet" bulb and other a "dry" bulb.

Temperature
is a measure of the degree or intensity of heat or cold.

Humidity
is the amount of moisture in the air.

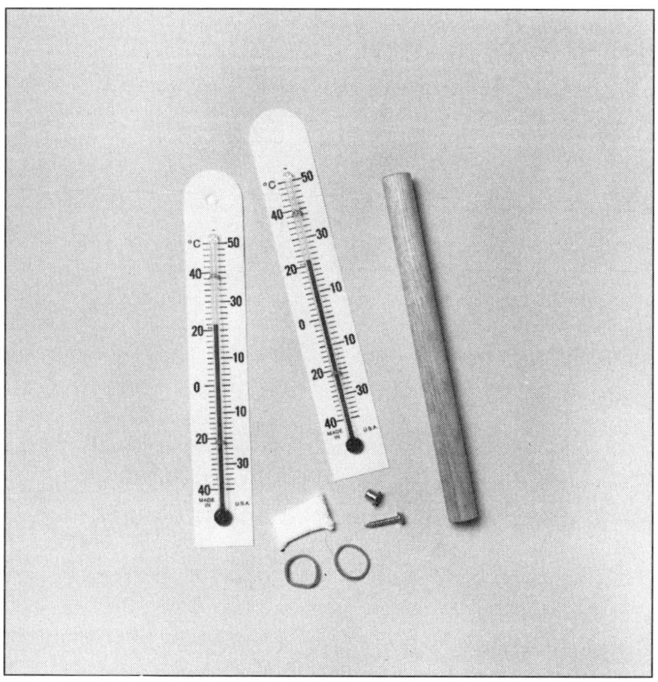

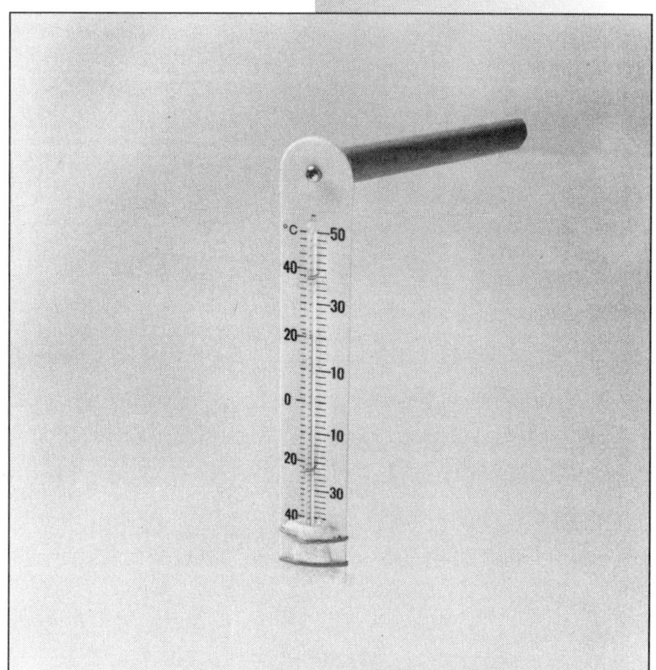

Student Psychrometer
To make a sling psychrometer: Attach two thermometers to a wood dowel with a screw. Fasten a piece of cloth to one of the thermometer bulbs with a rubber band. To use, dip the bulb with the cloth in water. Sling the thermometers around in the air for a few seconds. Read the wet-bulb temperature. Repeat until the wet-bulb temperature stops dropping. Read the dry-bulb temperature. Use the Relative Humidity Chart on page 64 to determine the relative humidity from the difference between the "wet" and "dry" temperatures.

Relative Humidity Chart

Dry-Bulb Reading (°C)	Difference Between Wet- and Dry-Bulb Readings (°C) 1	2	3	4	5
10	88	77	66	55	44
11	89	78	67	56	46
12	89	78	68	58	48
13	89	79	69	59	50
14	90	79	70	60	51
15	90	80	71	61	53
16	90	81	71	63	54
17	90	81	72	64	55
18	91	82	73	65	57
19	91	82	74	65	58
20	91	83	74	66	59
21	91	83	75	67	60
22	92	83	76	68	61
23	92	84	76	69	62
24	92	84	77	69	62
25	92	84	77	70	63
26	92	85	78	71	64
27	92	85	78	71	65
28	93	85	78	72	65
29	93	86	79	72	66
30	93	86	79	73	67

Cloud Cover: Determine the percentage of the sky that is covered with clouds as follows: 0–24%, 25%–49%, 50%–74%, 75%–100%. Simply eyeball the sky to estimate the percentage of cloud cover.

Sunshine: Have the students estimate the amount of sunshine using this verbal scale: sunny, mainly sunny, mainly cloudy, cloudy.

Wind: Estimate the speed of the air in km/hour and the wind direction.

Precipitation: Describe the type of precipitation (rain, snow, sleet, none), and the amount of precipitation in the last 24 hours in cm.

5. Have the materials managers get the equipment and a chart for their team. Each team will need two thermometers, rubber band, and small piece of cloth. Have the teams divide the responsibility for making observations among the team members. The tracker should record the data for the team. The checker should make sure that each person performs the following functions when they go outside to monitor the atmosphere:

Clouds
are visible masses of water or ice that have condensed on tiny particles in the air. Visit the Clean Air Web page in the GTP Web site to learn more about clouds, their formation, and types.

Wind
is the horizontal movement of air from higher to lower air pressure.

Precipitation
is moisture from the air in the form of rain, hail, sleet, snow, frost, or dew.

- Communicator—determines cloud cover and sunshine.
- Tracker—prepares chart and records data for the team, and measures precipitation.
- Checker—determines wind speed and direction.
- Materials Manager—uses psychrometer to determine temperature and humidity.

Data Collection

6. Take the students outside, and show them where to make the observations. Give them about ten minutes to make and record their observations.
7. After the students have returned from their first monitoring exercise, ask them to monitor the air each day during the week. Two persons from each team should be responsible for each day's readings. Make sure that everyone participates in the monitoring activity. Have each team post their data on a wall in the room and record their daily observations for all to see.

Data Transmission

8. Send a weather report to other schools in the GTP. Read the sample e-mail letter sent by Dobra Voda, Czech Republic Elementary. (See page 66.)

E-mail Message from the Dobra Voda, Czech Republic

Written 3:58 A.M., November 22, 1995 from "PaeDr. Jiri Vanicek" <vanicek@zf.jcu.cz>

————Electronic Weather Report ————

To: Global Community Teammates
Re: Weather Report for Dobra Voda, Czech Republic, Ceske Budejovice
Latitude: 48.52N Longitude: 14.32E

We made observations for five days, 13th November–17th November.
Here are the results:

	Day 1	Day 2	Day 3	Day 4	Day 5
Temperature	4	5	10	9	11
Humidity	100%	75%	77%	76%	67%
Cloud cover	75-100%	50-75%	25-50%	50-75%	0-25%
Sunshine	cloudy	cloudy	sunny	m. sunny	sunny
Wind speed	12-9km/h	1-5km/h	1-5km/h	6-11km/h	6-11km/h
Wind direction	E	NW	W	E	N
Precipitation	none	none	none	none	rain 0.8mm

Comments:

We measured temperature on the terrace in Celsius. Measure was above scale zero. (Small Tigers)

Humidity: We took two thermometers and one of these we covered the top and watered. We swung both about ten minutes on the terrace. We wrote down the temperature on the dry thermometer and the wet thermometer and then calculated the difference. (Chemikers)

Sunshine: We observed sunbeams this morning and discussed whether they were shining constantly or not. We found that clouds occasionally overlap the sun. (Teammakers)

Wind speed: We found that the wind was mostly strong, only on Wednesday, and Tuesday was it slight. We have the special paper, where we wrote how the wind blows (Beauford table of wind speed). (Greenpeace)

Direction of the wind: We used a plastic bag and it showed how the wind blew.

We measured cloudy weather by observation. We have observed if the sun is shining. We evaluated sunny, more sunny, more cloudy, cloudy. (Triangles)

Precipitation: Measured rain this way: We placed the bowl on the terrace near our school. When it started to rain, the bowl filled with rainwater. When we poured rainwater into a measuring bowl we determined the 18th November 95 share of rain—0.8ml. Rainwater was measured only this day because the rest of the week it did not rain. (Globics)

****

7th grade students
Dobra Voda Elementary
Ceske Budejovice
Czech Republic

****

Post a short report on the bulletin board. You can summarize the week's weather by sending an electronic report like the one on page 66.

Using the Web

9. Have your students connect to the Project Clean Air home page to determine the weather conditions on at least three different continents. Make sure that if you are a Northern Hemisphere school, that the students investigate the weather in a Southern Hemisphere GTP site.

ACTIVITY 4.3

Exploration

Monitoring Particulate Matter

In this activity your students will investigate the health and environmental effects of particulate matter found in the air. *Particulate matter* is the term used for a mixture of solid particulates and liquid droplets found in the air. Coarse particles (larger than 2.5 micrometers) come from a variety of sources including windblown dust and grinding operations. Fine particles (less than 2.5 micrometers) often come from fuel combustion, power plants, and diesel buses and trucks. These fine particles are so small that thousands of them could fit on the period at the end of this sentence.

Suspended particulates are actually pieces of ash or smoke, soot, dust and liquid droplets released into the air by the burning of fuel, industrial processes, agricultural practices, and a number of natural processes. For example, volcanic eruptions can spew tons of ash particulates as high as 10–15 miles into the atmosphere. Winds aloft can carry these particulates around the planet, and depending on the amount and size of the particles, they can stay aloft for years, affecting sunsets and average world temperatures.

Particulates, when inhaled, can lodge in the lungs and contribute to respiratory disease by making breathing more difficult. For example, a study reported at an international conference of the American Lung Association found that tiny airborne particles help to kill many people weakened by old age, heart problems, or lung disease. The study concluded that as many as 3 percent of U.S. deaths are attributable to suspended particulate matter, despite particulate levels in most urban areas that are well below the U.S. Environmental Protection Agency (EPA) health standard—a limit of 150 micrograms per day.[2]

[2] The U.S. Environmental Protection Agency Web site (See Resources.)

Some Frequently Asked Questions About Particulates

1. **What is *particulate matter* and what are "fine" particles?**
   Particulate matter comes from a variety of sources, including diesel trucks, power plants, wood stoves, and industrial processes. While individual particles cannot be seen with the naked eye, collectively they can appear as black soot, dust clouds, or gray hazes. Those particles that are less than 2.5 micrometers in diameter are known as "fine" particles; those larger are called "coarse" particles. Fine particles can be formed in the air from gases; coarse particles are usually emitted from sources such as vehicles traveling on unpaved roads, materials handling, crushing and grinding operations, and windblown dust.
2. **Why are we concerned about particulate matter?**
   Particulate matter is the term for a mixture of solid particles and liquid droplets found in the air. These particles can be very small. For example, fine particles (less than 2.5 micrometers) can reach the deepest recesses of the lungs. According to the EPA, many scientific studies have linked particulate matter, especially the fine particles, with a series of health problems, including:
   - Respiratory-related hospital admissions
   - Aggravated asthma
   - Acute respiratory symptoms, including aggravated coughing and difficulty breathing
   - Chronic bronchitis
   - Decreased lung function, which can be experienced as shortness of breath
   - Work and school absences
   - Premature death
3. **Who is most at risk from exposure to fine particles?**
   The elderly, individuals with pre-existing heart or lung disease, children, and asthmatics. Studies indicate that many elderly people die prematurely every year from exposure to fine particles. Breathing fine particles can also adversely affect individuals with heart disease, emphysema, and chronic bronchitis. The average adult breathes 13,000 liters of air per day; children breathe 50 percent more air per pound of body weight than adults. Because children's respiratory systems are still developing, they are more susceptible to environmental threats.

4. **How do particulate matter and fine particles visibly affect the environment?**
Fine particles that are linked to serious health effects are also a major cause of visibility impairment. For example, in many parts of the United States the visual range has been reduced 70 percent from natural conditions. In the East, the current range is only 14–24 miles vs. a natural visibility of 90 miles. In the West, the current range is 33–90 miles vs. a natural visibility of 140 miles. Fine particles can remain suspended in the air and travel long distances. For example, the EPA reports that a puff of exhaust from a diesel truck in Los Angeles can end up over the Grand Canyon in Arizona, where one-third of the haze comes from Southern California.
5. **What improvements would result if we met the new U.S. Environmental Protection Agency standards?**
The new particulate standards should provide the following benefits:
   - About 15,000 lives will be saved per year, especially among the elderly
   - Reduced risk of hospital admissions by thousands each year
   - Reduced risk of symptoms associated with chronic bronchitis
   - Reduced risk of respiratory symptoms in children
   - Reduced risk of aggravation of asthma

   The new standard should also lead to improved visibility, especially over urban areas.

   In the activity below, students will learn to monitor particulates in their own environment (indoors or outside) by making a piece of equipment from ordinary materials and then collecting samples of particulates from different locations for comparisons.

Objectives

- Design a device that can be used to monitor the quantity of particulate matter in the atmosphere.
- Detect and measure particulates in the local environment.
- Communicate the results with others in the GTP.

Materials

file folders, card stock or 10 cm × 15 cm index cards, clear sticky tape, plastic margarine containers, balance, large glass jar with a cover, paper, matches, Student Log pages 209, 210

A simple demonstration such as burning a piece of paper in a jar can illustrate air pollution.

Procedures

1. Start the lesson with a brief demonstration. Ignite a piece of paper and drop it into the large glass jar. Cover the jar. "Smoke" will fill the jar. Ask the students how the "atmosphere" inside the jar changed as a result of the burning paper. Students will respond with statements such as: "The smoke made the air dirty." "Carbon particles filled the air in the jar." Help the students realize that particles from the burning process were released into the air in the jar. These particulates are air pollutants. Ask the students to name some sources of particulate air pollutants in the atmosphere. Students may mention smoke from burning wood, dust from soil, pollen, windblown sand, volcanic ash, soot from factories, and so forth.
2. Point out that particulates cause harm to humans. Show your students the graph below that illustrates the amount of suspended particulates put into the atmosphere just in the United States from the period 1970–1989. Point out that particulates are one type of air pollutant. Air pollutants can be solid, liquid, or gas. Particulates are solid pollutants. An example of a liquid pollutant is acid rain (which will be explored in another project). Examples of gas pollutants are carbon monoxide, sulfur dioxide, nitrogen oxides, and ozone. Ozone will be explored in the next activity.

U.S. Total Suspended Particulates, 1970–1989[3]

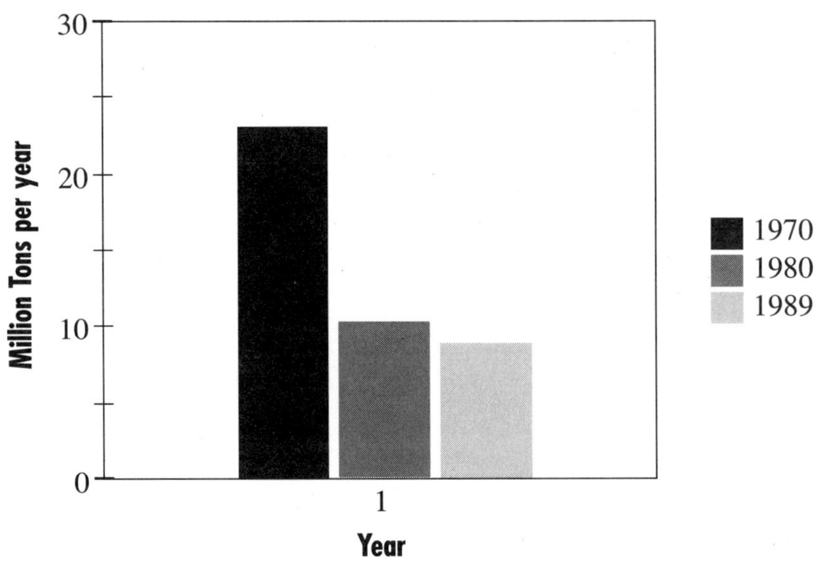

[3] 1992 Information Please®, Environmental Almanac, World Resources Institute.

3. Give students materials to make a device to monitor atmospheric particulates.

Design

1. Give each team two or three copies of the Air Quality Monitoring Card, clear plastic tape, scissors, and string.
2. Have the students cut out squares (2.54 cm) as shown on the card. Cover each square "window" with a piece of clear plastic sticky tape. Make sure the sticky side is placed on the underside of the card so the sticky side is facing out.
3. Punch a hole in the top of the card and fasten a piece of string through the hole. The card can be suspended from or taped to objects for monitoring (e.g., buildings, windows, trees).
4. Fill in the data, location, and team name, and then carefully place the card in the desired location.

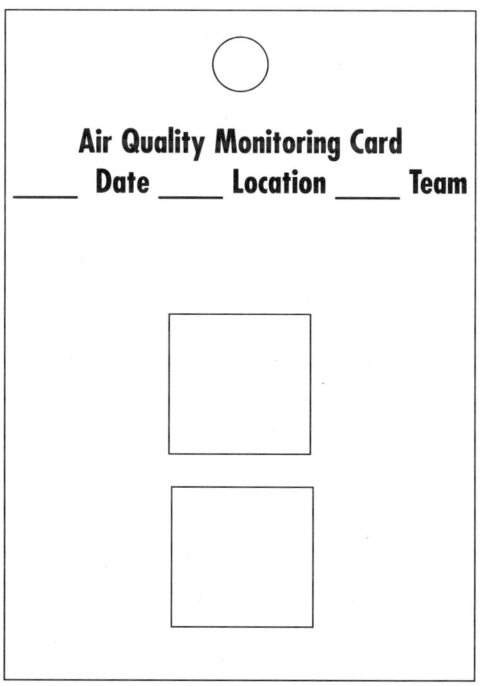

Air Quality Monitoring Card
Use this design as a template to make copies for your students.

Methods of Use

Hang the cards in different locations for seven days. Photocopy mm graph paper onto an overhead transparency. Cut pieces of these grid transparencies, and place them on the sticky side of the windows on the particulate cards at the end of the seven-day exposure period. Examine the collectors under a microscope, or with a magnifying glass. Estimate the number of particles per mm^2, and record the shapes and colors of the particles.

Record Data

Using the following data form, have students record their data.

Outdoor Particulate Data Form

Team Name: ______________ Team Members: ______________

Measurement Location: Inside __________ Outside __________

Starting Date: ______ Starting Time: ______ Latitude: ________

Ending Date: ______ Ending Time: ______ Longitude: ______

Method Used: Sticky Tape Cards _________ Other: __________

Observations

Particulate Count: High Moderate Low

Particulate Density
(match shading in window using this shading scale)

1	2	3	4	5

Type of Particulates: ______________________________________

Size of Particulates:
(% in each category) ____ <0.1 mm ____ 0.1–1.0 mm ____>1.0 mm

Other Observations: ______________________________________

Data Analysis and Conclusions

Have the teams analyze their data and the effectiveness of the method they used. Were there any problems that they encountered? What were the problems? How can they change their methodology to improve their results?

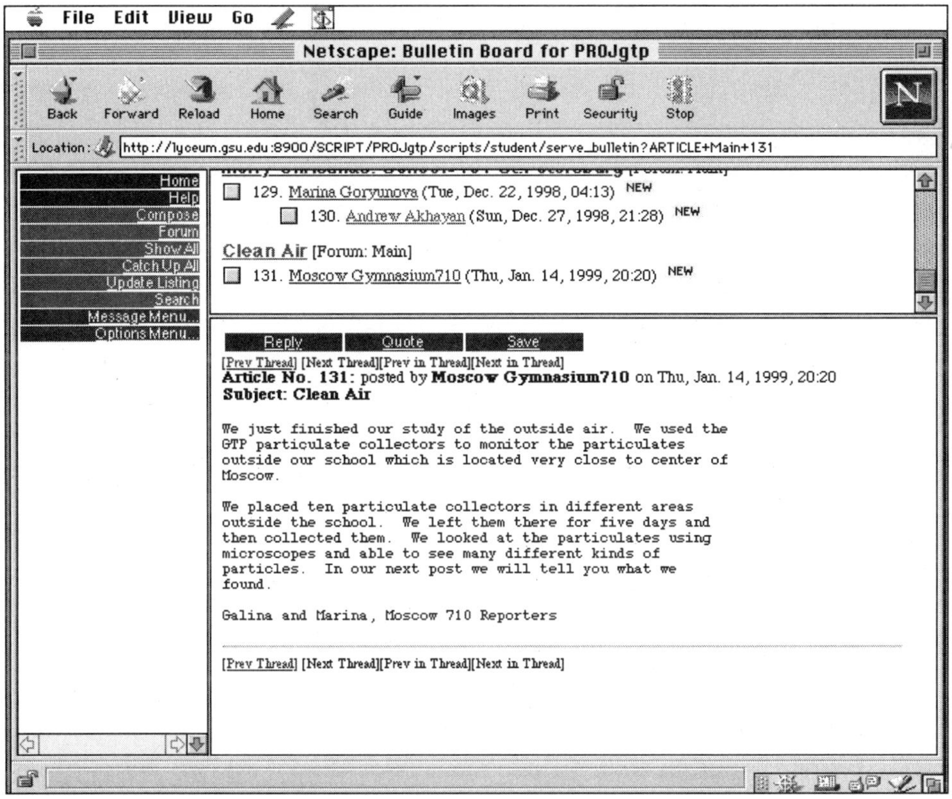

Bulletin Board Report Sample

Have the class summarize the results of their work, and then share the results with the GTP by sending a report to the Web Discussion for Project Clean Air. Encourage the students to formulate brief messages for this activity. In the case of the Air Quality Monitoring Card, they might report their results in terms of the distribution and types of particles, as well as include a statement evaluating the effectiveness of their method and any suggestions that might help other student teams use this method.

ACTIVITY 4.4

Monitoring Ground-Level Ozone

Exploration

Most of your students have probably heard of smog. Smog forms when primary pollutants such as nitrogen oxides (NO_x), which are produced by burning fuels, and volatile organic compounds (VOCs) escape into the atmosphere. VOCs are unburned hydrocarbons produced by automobile fuel tanks, other chemicals from commercial enterprises, and natural sources such as trees. In the air, these chemicals react or combine with sunlight to produce a number of different materials, the most common of which is ozone (O_3). Because of this, ozone is an example of a photochemical smog. This form of ozone is concentrated in the lower atmosphere (the troposphere), where it comes in contact with plants and humans. High levels of ozone can result in eye irritation and respiratory

Smog
is a dense, discolored fog containing large quantities of soot, ash, and gaseous pollutants.

Ozone
is a highly reactive gas comprising three oxygen atoms. Ozone is produced by the recombination of oxygen in the presence of ultraviolet radiation from the sun. Ground-level ozone is an atmospheric pollutant that is formed by the reaction of hydrocarbons and nitrogen oxides in the presence of sunlight.

problems. Ozone can also damage plant leaves and slow growth. In this activity, students measure ozone levels in and around their school using a chemically treated paper, called Eco™Filter. City smog is a serious global problem, and one worthy of investigation.

When students hear the word *ozone,* they may refer to the "ozone hole." The ozone hole refers to ozone located in the stratosphere that protects us from the ultraviolet rays of sunlight. Although students are not going to investigate the good ozone in this project, they will want to talk about it. You might tell them that certain chemicals, such as chlorofluorocarbons (CFCs) used in refrigerators, air conditioners, aerosol sprays, cleansers for electrical parts, float to the stratosphere and photochemically react with ozone. The CFCs react directly with ultraviolet light releasing chlorine. The chlorine atom reacts with and destroys the O_3. The destruction of the ozone layer is considered to be one of the most serious global problems of the day.

Frequently Asked Questions About Ground-Level Ozone[4]

1. **What is ground-level ozone?**
   Ozone is a gas, O_3, that is not emitted directly into the air. Instead it is formed by gases called nitrogen oxides (NO_x) and volatile organic compounds (VOCs) in the presence of heat and sunlight. This reaction occurs readily in the atmosphere, especially on hot sunny days in urban areas.
2. **Why are we concerned about ground-level ozone?**
   When inhaled, even at low levels, ozone (which is the prime ingredient of smog) can:

   - Cause acute repiratory problems
   - Aggravate asthma
   - Cause significant temporary decreases (up to 20 percent) in lung capacity
   - Cause inflammation of lung tissue
   - Lead to hospital admissions and emergency room visits (10–20 percent of all summertime respiratory-related hospital visits in the northeastern part of the United States are associated with ozone pollution)
   - Impair the body's immune system defenses, making people susceptible to bronchitis and pneumonia

[4] Adapted from the EPA Web site (See Resources.)

3. **Who is most at risk from exposure to ozone?**
Children, the elderly, people with asthma, and adults who exercise outdoors are most at risk to ozone exposure. Children breathe more air per pound of body weight than adults and their respiratory systems are still developing. They also tend to be outside playing (especially in the summer). Surprisingly, healthy adults, those that are exercising outdoors, are also at risk.
4. **How does ground-level ozone harm the environment?**
Ground-level ozone interferes with the ability of plants to produce and store food. Ozone can make plants more susceptible to disease, pests, and environmental stress. Ground-level ozone can kill or damage leaves so that they fall off plants too soon or become spotted or brown.
5. **What improvements would result from the EPA's new standards?**
Increasing the ozone standard should reduce the risk of significant diseases in children's lung functions. It is estimated that a million fewer incidences per year would result. There would also be a reduced risk of moderate to severe respiratory symptoms in children, resulting in hundreds of thousands of fewer incidences. There should be fewer hospital and emergency room visits for respiratory causes. A reduced yield loss of major agricultural crops, such as soybeans and wheat, and commercial forests should result.
6. **What is the ozone hole and how is it related to ground-level ozone?**
Ozone gas exists in the upper part of Earth's atmosphere. Ozone located in the upper atmosphere forms a layer of gas that protects living things on Earth from the harmful effects of ultraviolet light. Human-made gases, especially CFCs, have drifted into the upper atmosphere where they have "attacked" ozone, reducing the amount of ozone in the protective layer. Over some parts of Earth, especially in the Southern Hemisphere, "holes" have developed in the ozone layer. The ozone hole, therefore, is not related to ground-level ozone.

Objectives

- Measure ozone levels in the local environment.
- Compare ozone measurements with students from other locations.
- Communicate results with others in the GTP.
- Share data using a World Wide Web browser form in the GTP Web site.

Materials

Eco™Filters[5], ozone test strips, graph paper, thermometers, psychrometers, Student Log pages 211, 212

Procedures

Ozone Observations and Measurements

1. Explain to students that they are going to investigate another type of pollutant, one that is produced when primary pollutants combine in the atmosphere to produce a secondary pollutant. To start off, ask students what effect a "Smog Alert" would have on the activities of people. What cities might be likely to have smog alerts? Why? What effect do they think temperature has on smog? Discuss smog with the students, and point out that the main component of smog is ozone. You might want to use the introductory material in this project to give a short presentation on ozone. Follow this by having student pairs draw a diagram showing how tropospheric ozone differs from stratospheric ozone.
2. Show the students an Eco™Filter, a small strip of filter paper that has been treated with two chemicals—one that is sensitive to ozone so that in one hour it turns a shade of violet, and another that is sensitive over an eight-hour period and turns a shade of brown. We will refer to the Eco™Filter as an ozone test strip. Tell them that they will be using the ozone test strips to measure ozone.
3. Describe how the ozone test strips are used. Tape the ozone test strips on a small index card. Since the chemical works when it is exposed to the air, the students must work quickly. If you are going to do a one-hour test, you should cut the strips and give students only the top part of the strip (the one-hour test circle). This should be taped to the card.

   Another technique is to cover the test strip with a piece of clear acetate. Cut small holes the size of the test sections and tape the acetate directly over the ozone test strip. Cover the one-hour test section with one of the circles of acetate. You can then run an eight-hour test, and then do a one-hour test at another time.

   When students place the card in the monitoring location, they should observe the temperature, humidity, cloud cover, cloud type, visibility, and weather. Provide data tables for students to record their data.

[5] Order from Vistanomics, Inc. (See Resources for ordering information.)

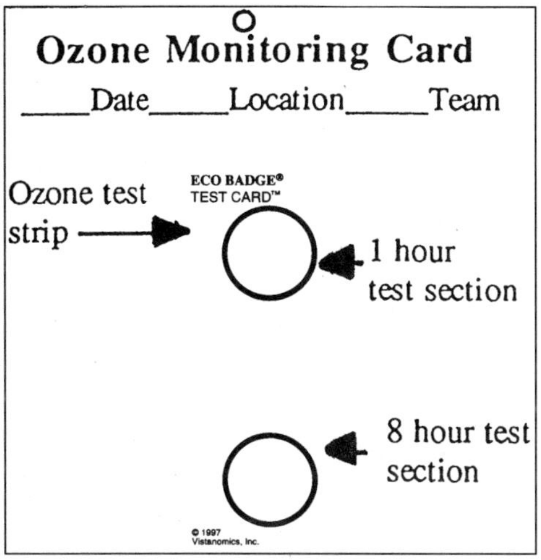

Ozone Test Strip and Card

4. The students can monitor ozone on school grounds for a two-day period. They should record their observations on the Project Clean Air Recording Form (Student Log page 212). Be sure that students record quantitative and qualitative data each day. The observations should be recorded on a class chart and in a data table in the computer.

Project Clean Air Recording Form Sample

Team Name: ____________ Team Members: ____________

Measurement Location: Inside ____________ Outside ____________

Date: ____________ Latitude: ____________

Start Time: ____________ Longitude: ____________

Quantitative Measurements

Ozone (Eco Badge® Readings)

Location	Start Time	End Time	Reading 1 ppb	Reading 2 ppb	Reading 3 ppb	Average ppb

Temperature: ____________ °C

Relative Humidity: ____________ %

Wind Speed: ____________ km/hr

Wind Direction: ____________ degrees

Precipitation: ____________ cm

Qualitative Observations

Circle one of the words in each row to describe your observations.

Ozone	Good	Moderate	Unhealthful	1st Stage Alert
Temperature	Hot	Warm	Cool	Cold
Humidity	High	Average	Low	
Clouds	Cloudy	Partly cloudy	No clouds	
Sunshine	Sunny	Mainly sunny	Mainly clouds	Cloudy
Wind Speed	Strong	Medium	Light	Zero
Wind Direction	North	East	West	South
Precipitation	Rain	Drizzle	Snow/sleet	None
Particulates	High	Moderate	Low	

Other Observations: ____________

____

____

Ozone Data Class Chart Sample

Observations	**Day 1 (date)**	**Day 2 (date)**	**Other Schools' Data (make additional columns)**
Quantitative Data			
Ozone			
Temperature			
Wind Speed			
Wind Direction			
Precipitation			
Latitude			
Longitude			
Qualitative Data			
Ozone			
Temperature			
Humidity			
Clouds			
Sunshine			
Wind Speed			
Wind Direction			
Precipitation			
Particulate Count			

Interpreting Ozone Observations

To read the strips, the color on the ozone test strip should be compared to the Eco Badge®[5] Colormetric Chart.

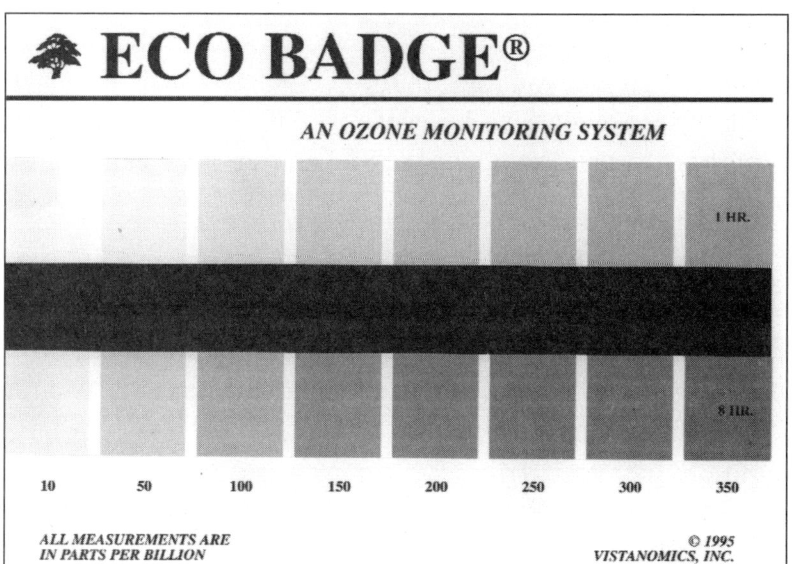

EcoBadge Chart Sample

5 Reprinted by permission of Vistanomics, Inc.

To interpret the meaning of the ozone level, share the data in the ozone conversion table below. The U.S. EPA has established 120 parts per billion of ozone as the optimal level allowable by law. That is, if a reading exceeds 120 ppb (even for one day), then that community is not in compliance. Naturally, some cities have a more serious problem than others. For example, these cities were above the smog level standards an average of the following number of days per year: Los Angeles (155); New York (20); Houston (20); Atlanta (5); Fresno, CA (21); and Bakersfield, CA (29).

Ozone Conversion Table

Air Quality	Good	Moderate	Unhealthful	1st Stage Alert	2nd Stage Alert
PPB on Eco™Filter	10–50	50–150	150–250	250–350	>350
Part per billion	<60	70–120	120–190	190–340	>340
Ozone Index reported to public. (ozone level/ 120 × 100)	50	58–100	100–158	158–280	>280

Using the World Wide Web

Data Transmission

After the data have been compiled, the class should submit their data to the Project Clean Air database. From the Project Clean Air home page, select "Data Sharing." Students will be able to enter their data directly into a Web form (See the data form on page 81.) and send it electronically to the database.

Data Retrieval and Analysis

1. From the Project Clean Air home page, select "Data Retrieval." The data submitted by other schools will be shown as a data table. You can download this data directly to your computer, or use the data table directly from the Internet.

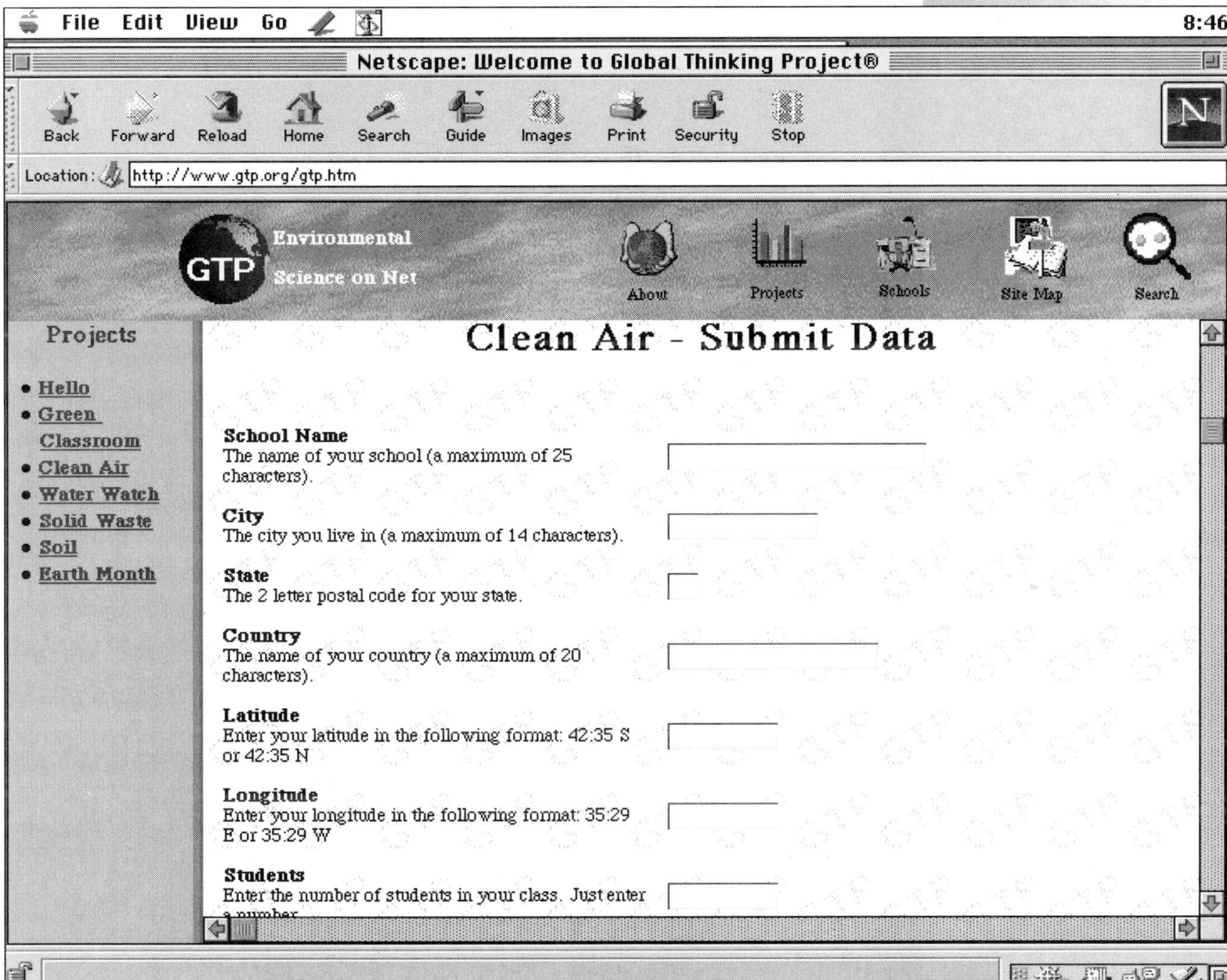

Project Clean Air Submit Data Form

2. Have your students look over the data. Have them write at least five questions they could answer using the data they have available.
3. Have each team select one question and prepare a report for the class in which they state the question, present the relevant data, and draw conclusions based on the data. Give teams time to use the computer to make graphs, charts, and maps of the data. Have students prepare a poster report to be presented to the rest of the class.

Web Discussion

From the Project Clean Air home page, select "Web Discussion." Post the results of the student reports in discussion format. Respond to other postings with your teams' comments and questions.

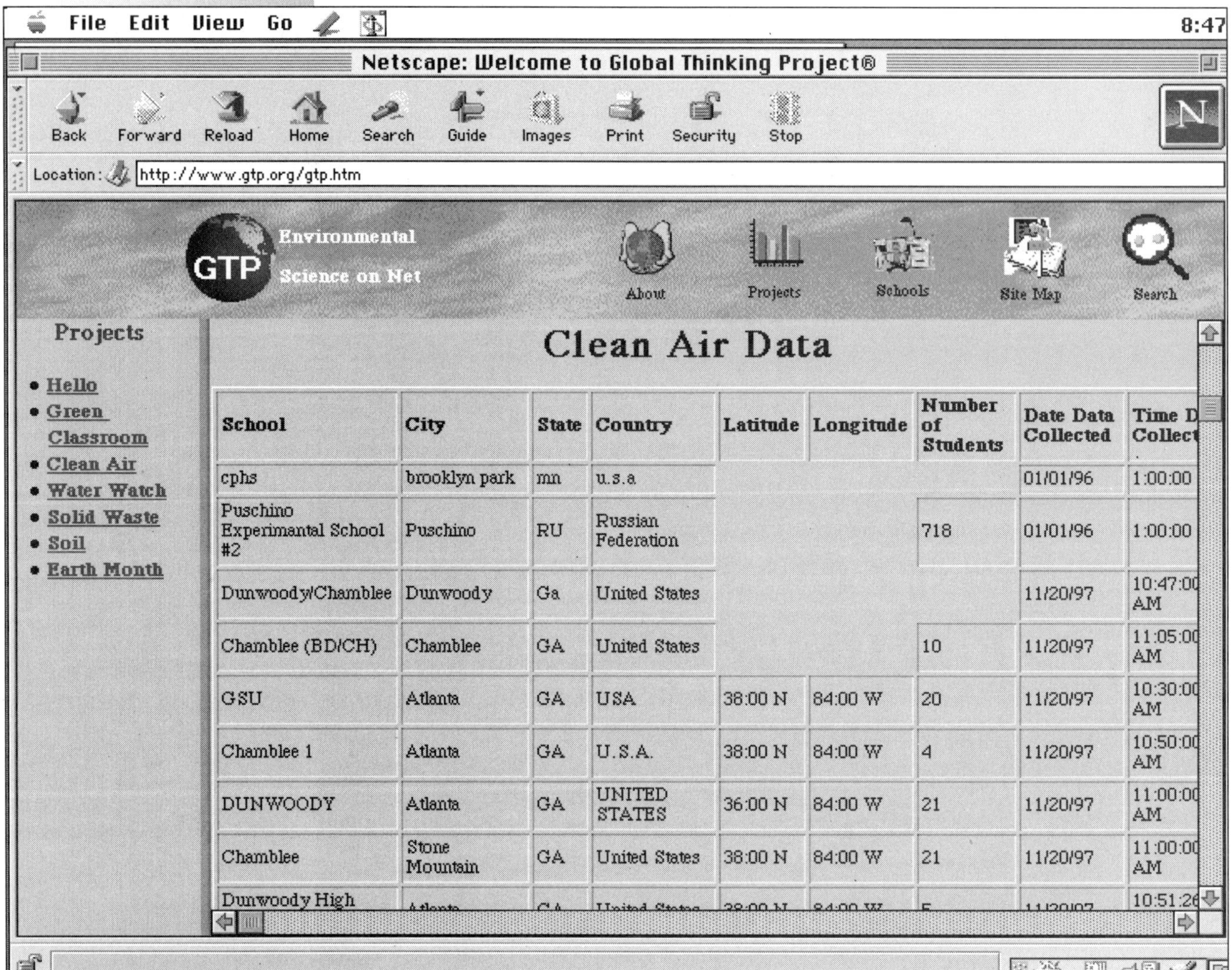

School	City	State	Country	Latitude	Longitude	Number of Students	Date Data Collected	Time D Collect
cphs	brooklyn park	mn	u.s.a				01/01/96	1:00:00
Puschino Experimantal School #2	Puschino	RU	Russian Federation			718	01/01/96	1:00:00
Dunwoody/Chamblee	Dunwoody	Ga	United States				11/20/97	10:47:00 AM
Chamblee (BD/CH)	Chamblee	GA	United States			10	11/20/97	11:05:00 AM
GSU	Atlanta	GA	USA	38:00 N	84:00 W	20	11/20/97	10:30:00 AM
Chamblee 1	Atlanta	GA	U.S.A.	38:00 N	84:00 W	4	11/20/97	10:50:00 AM
DUNWOODY	Atlanta	GA	UNITED STATES	36:00 N	84:00 W	21	11/20/97	11:00:00 AM
Chamblee	Stone Mountain	GA	United States	38:00 N	84:00 W	21	11/20/97	11:00:00 AM
Dunwoody High								10:51:26

Project Clean Air Data Table

ACTIVITY 4.5

Clean Air Student Projects

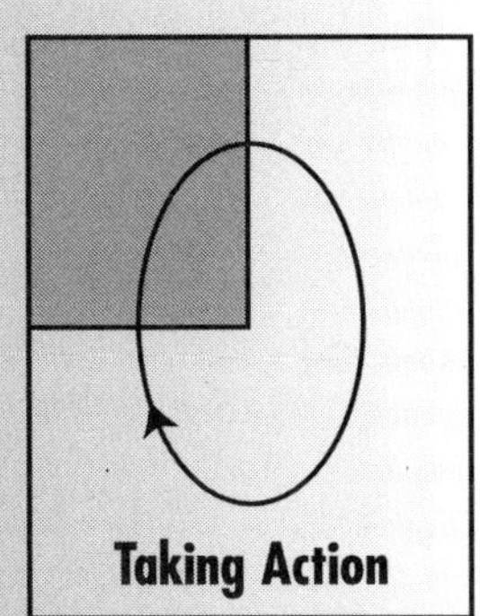

The purpose of this activity is for students to design "Clean Air" projects.[6] Student should work in cooperative teams to design and carryout their projects. Students will be involved in the following procedures:

- brainstorming questions that they would like to investigate about air pollution
- designing a research method or set of procedures to help them answer their question(s)
- monitoring the variables that are being investigated in their study
- analyzing data including both graphical and verbal analysis
- summarizing results and evaluating their methodology
- engaging others via the telecommunications network; this can simply be the reporting of their research plan and results, or it can go as far as designing project that depends on the participation of students in remote classrooms
- responding to network invitations to collaborate with other student research teams

Objectives

- Design projects based on questions and inquiries about air quality
- Identify, discuss and use research skills to answer scientific questions
- Use computer-based technologies to collaborate with students in the Global Thinking Project.

Materials

Student Log page 213; materials will vary from project to project

Procedures

1. Begin the activity by writing the phrase: *Act Locally, Think Globally* on a large sheet of paper or on the chalkboard. Tell the students that they are going to "act locally" by doing a project that focuses on air pollution in the local environment. However, because they are part of the telecommunications network that brings them in touch with students around the world, their study can have a global dimension.

[6] You might want to postpone this activity and use it as a Project Earthmonth investigation for your class. During the Project Earthmonth, your students will be asked to design their own environmental projects. This might serve as an excellent outline for a project.

2. Discuss with your students the nature of a research project. You might begin by posing a question to your students, and then asking them how such a question might be answered.
   - Are there more particulates in the air on windy days than on days when there is just a slight breeze? or
   - Is the ozone level near plants (such trees and shrubs) different than the ozone level in an open field?

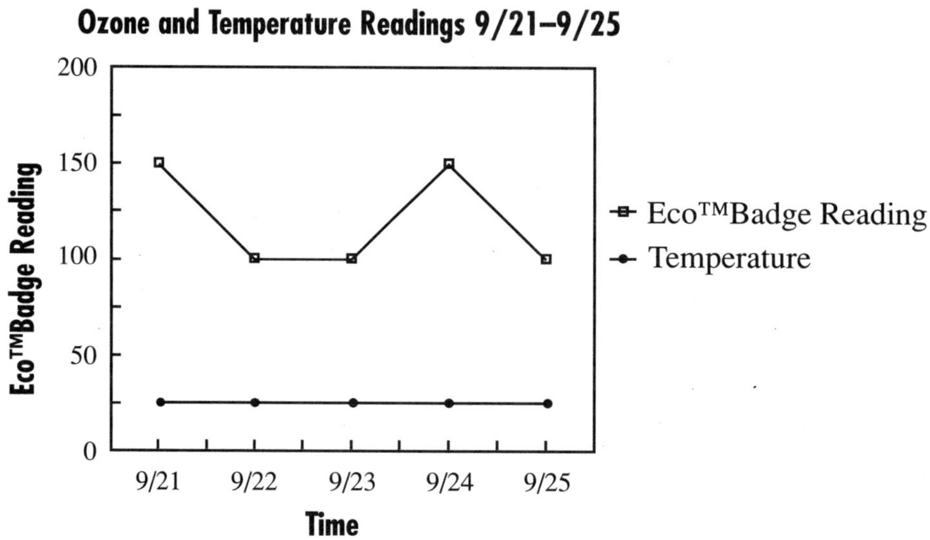

Sample data from a study investigating the relationship between ozone levels and temperature.

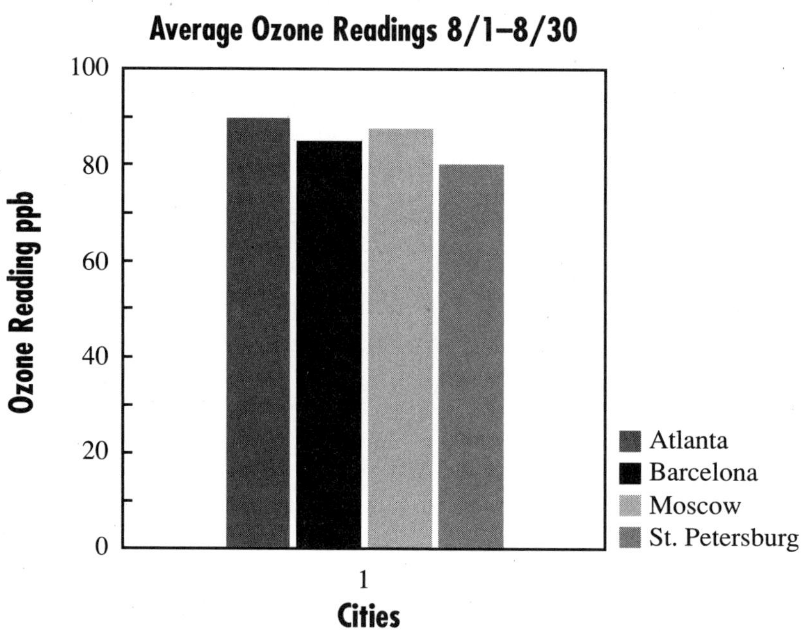

Graph of data collected by GTP students in four different cities.

Give each team an assignment to design a study that would answer one of the questions. Tell the students simply to list the steps (or the method) they would use to answer the question they have been given. When the students are finished, ask a representative from each group to describe the method of its study.

3. Use the procedures in the Student Guide for Outdoor Air Projects on page 86 to guide the students as they do their project.

Student Guide for Outdoor Air Projects

Project Step	Student Activity
Step 1: Brainstorm Questions	Brainstorm a set of questions and agree on a focus for the study. Students should reflect on the work they have done so far on air quality and suggest to each other questions about atmospheric monitoring, particulates, and ozone. You might have them look at the initial list of questions they wrote in Activity 4.1 and see if there are any they would like to tackle.
Step 2: Agree on Questions	Using consensus, students agree on the research question(s) they wish to investigate. Then they should write a brief statement describing the focus of their study.
Step 3: Describe the Study Plan	Students should identify the variables that will be important to monitor in the study. Given the variables and the research questions, the students should describe how they plan to carry out their study. The plan can be an outline of the steps they will follow in their study.
Step 4: Check the Network	At this stage, you may want teams to post the description of their study on the Project Clean Air home page, especially if they are doing a study that seeks collaboration from other cities or countries. It is important to note that if your students are seeking collaboration from others, you must take the responsibility to check the network daily to: • Identify teams of students who do indeed wish to work with your students • Respond to students in other schools who are looking for research partners
Step 5: The Study	Students should conduct the study, which may involve contacting other students on the telecommunications network and asking for help in collecting data at remote sites. It may also mean going to monitoring sites during the day, or taking equipment home to collect data at night or on the weekends. During this step, students will be involved in a number of activities: • Monitoring the variables in their study • Visiting the Project Clean Air home page for resources • Writing letters or making phone calls to organizations for information about the topic
Step 6: Analyze the Data	Students should organize their information in a systematic manner. This might involve the use of data tables, graphic analysis, and discussions within small teams and with the teacher. Some teams may need to collaborate with peers via the network. The results of the study should be analyzed in preparation for the development of a final report. Note: if your students collaborated with another group, make sure they send the data to other team(s).
Step 7: Prepare Reports	Student teams should prepare a poster report of their project and write a report that can be posted on the discussion in the Project Clean Air home page.

4. When students have completed their project, they should prepare poster reports for class presentations and send reports to share to the GTP bulletin board.

Web Discussion

From the Project Clean Air home page, select "Web Discussion." Have students submit brief reports of their research projects. If they developed a Web page for their project, have them post their Web site address so that other GTPers can browse through it.

Optional Extensions for Project Clean Air

1. Have the students write an editorial letter or report of their study, and the implications for the local community, and send it to the local newspaper.
2. Have the students create a scrapbook of articles on pollution (don't limit them to air pollution).
3. Make contact with a local environmental action group and let them know that they are working on global problems. Students can send the local group a copy of their report, and encourage the group to visit their class.
4. Start a Global Thinking Ecology Club. Students can meet to discuss the idea, then create a charter, and begin a series of activities focusing on the local environment.
5. Have students create a Web site that focuses on the problems and solutions related to air pollution.

CHAPTER 5

Project Solid Waste

Every year we create more waste.
Every year there's less space to put it.
The *1992 Information Please® Environmental Almanac*

Landfill space is rapidly disappearing, and cities all over the world are having to find new solutions to the problem of where to put their trash. In 1997 the city of Atlanta, Georgia, was ordered by the state environmental protection division to close its two major landfills after the state lowered the amount of trash they were allowed to accept. Attempts to extend the life of existing landfills in some locales include banning yard trimmings from landfills and encouraging homeowners to reduce their garbage output by charging by weight for its collection. In Project Solid Waste and the activities in this chapter, students examine their own personal waste streams and begin to identify ways that they might contribute to a reduction in the production of nonbiodegradable solid waste. (See the planning chart on page 90 to see how time for each activity should be allotted.)

As landfill space becomes limited, future generations will have to discover new ways to reduce garbage output.

Goals for Project Solid Waste

1. Students will become familiar with the terms and concepts associated with disposal of nontoxic solid waste.
2. Students will explore the extent of the solid-waste disposal problem on a local and global level.
3. Students will identify and evaluate various means of disposal.
4. Students will suggest viable solutions to the disposal problem.

5. Students will network with schools in other communities. Networking activities will include sending other schools' data on the composition of the waste stream, analyzing other schools, solid-waste data, and engaging in collaborative projects.

Planning Chart

Activity		Class Periods
5.1	Getting to know your trash	1–2
5.2	The 3Rs: Reduce, Reuse, Recycle	1–2
5.3	How long will it be there?	1–2, with weekly checks
5.4	Evaluating waste-disposal options	2–3

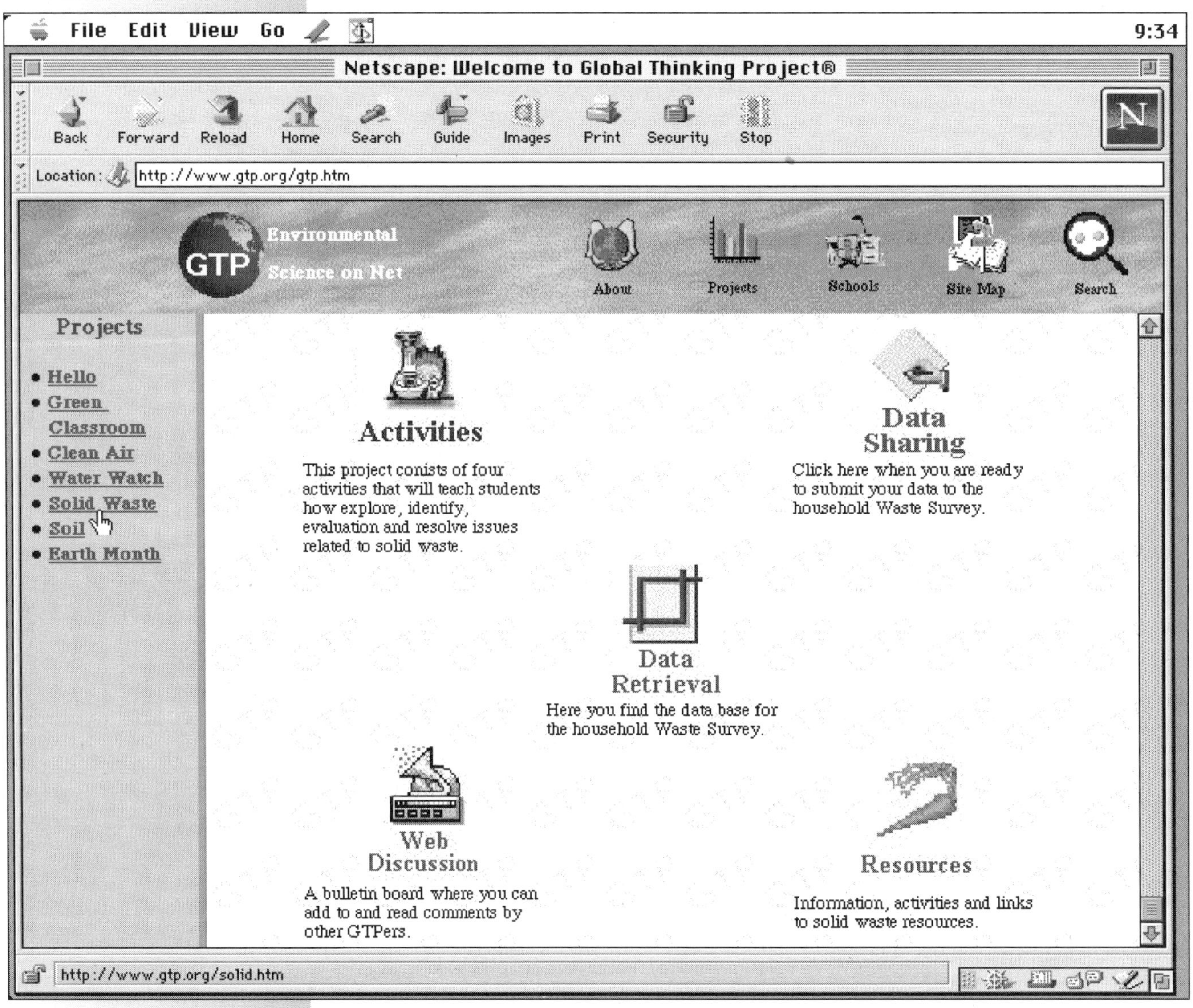

Project Solid Waste Home Page

Web Site Information

From GTP home page (www.gtp.org) click on "projects" to access Project Solid Waste home page.

ACTIVITY 5.1

Getting to Know Your Trash

Invitation

In this activity, students classify and measure the solid-waste materials produced by their families.

Objectives

- Develop a classification system for the solid waste produced by a family.
- Classify and measure the solid-waste output of a family.
- Share results of a solid-waste survey with students in other communities.

Materials

flip chart or chalkboard, markers, Student Log pages 214, 215

Procedures

1. Begin by asking students to name the things they threw out after lunch that day (list on board or chart paper). Ask those students who brought their lunches from home to add to the list things that were thrown out in the process of preparing their lunches.
2. Assist the students in classifying the different kinds of trash. What other kinds of things do we throw out at home? (Add other types of trash to the list.)
3. Ask students in teams to generate a list of questions they have about household solid waste. Post each team's questions on chart paper around the room.
4. Conduct a classwide investigation of household solid-waste output using the Household Waste Survey on page 214 in the Student Log.
    - For each category, students should record the weight of garbage they collected. Add to get the total daily garbage output.
    - Have them compute the percent of the total weight contributed by each category. Students will need to separate and weigh (on a bathroom scale at home) their family's output of different waste categories for one day.
5. Ask teams to compile their individual data to determine an average daily output per person for each category. Teams may present these findings in charts, tables, or graphs.

Household Waste Survey Sample

Category	Weight of Garbage	% by Weight
Paper and paperboard		
Metals		
Glass		
Plastics		
Rubber and leather		
Textiles		
Yard wastes		
Food wastes		
Other		
Totals		

Optional Extensions

1. Invite teams to explore other questions they have about the composition of household solid waste. Additional questions which may be explored by teams may include the following:

   - What is the average yearly output per person?
   - Based on these data, what would you estimate the average yearly output of your city or town to be?
   - What volume of space would one person's yearly solid-waste output occupy in a landfill? the city's?
   - Based on predicted population figures for your city, project the solid-waste output ten years from now, assuming waste continues to be produced at the same rate.

   Provide class time for teams to present their findings to each other.
2. According to the *1992 Information Please® Environmental Almanac,* the average daily solid waste generated per person in the United States in 1988 was 4 pounds. Of this, approximately 1.5 pounds were paper, 0.25 pounds were glass, and 0.30 pounds were plastics (with the remainder in other categories). How does your class data compare with these data? If your data are significantly different, speculate why this might have occurred.
3. Exchange survey results with students in other communities by posting a message to the Project Solid Waste Web discussion. How do the results compare? How do your results compare to the Waste in Major Cities graph on page 93?
4. Observe your neighborhood for evidence of solid-waste disposal problems. Inventory the trash in a small area, or create a photographic montage of clean and littered areas in your neighborhood. Share your findings with civic leaders.

Waste in Major Cities[1]

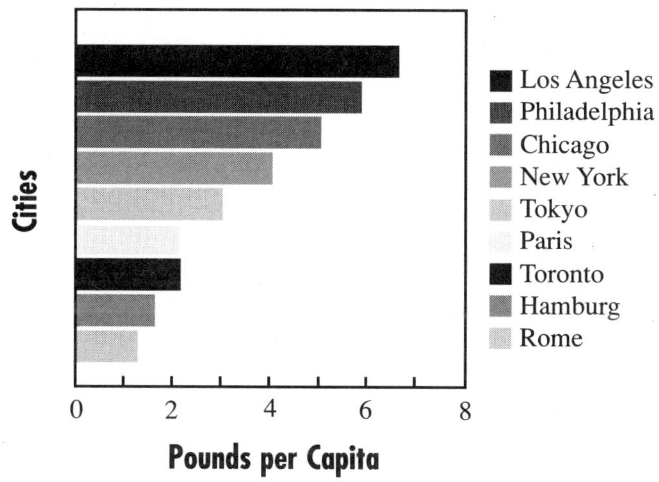

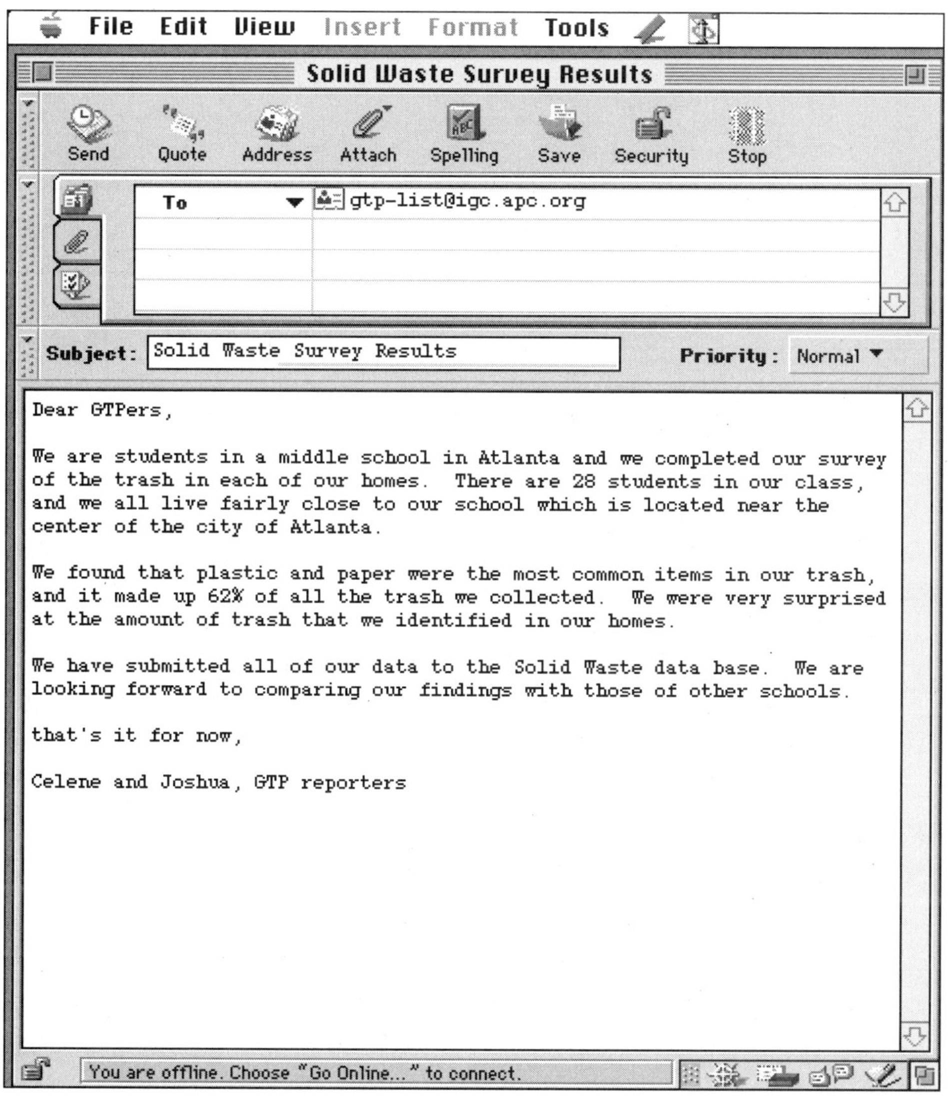

Dear GTPers,

We are students in a middle school in Atlanta and we completed our survey of the trash in each of our homes. There are 28 students in our class, and we all live fairly close to our school which is located near the center of the city of Atlanta.

We found that plastic and paper were the most common items in our trash, and it made up 62% of all the trash we collected. We were very surprised at the amount of trash that we identified in our homes.

We have submitted all of our data to the Solid Waste data base. We are looking forward to comparing our findings with those of other schools.

that's it for now,

Celene and Joshua, GTP reporters

[1] Source: NSWMA

5. Using data gathered from reference books, make a map showing waste production and disposal patterns all over the world. Are there any patterns?

Exploration

ACTIVITY 5.2

The Three Rs: Reduce, Reuse, Recycle

In this activity students explore how the Three Rs of solid-waste disposal could help reduce their own family's solid-waste output.

Objective

- Identify ways to reduce the production of household solid waste

Materials

Materials will vary.

Procedures

1. Begin by asking the class whether they have heard the terms *reduce, reuse,* and *recycle.* Ask students to explain what they think each word means in relation to solid-waste disposal.
2. Have students look at the lists of things they threw away at lunch yesterday. Explain that the goal of this lesson is to discover ways to decrease the amount of trash they throw away. Spend about five minutes having students brainstorm ways they could reduce the trash from their lunches.
3. Assign one or two teams to conduct each of the investigations below and to report their findings to the class.
   - What's in your purse/pocket/desk/lunch box/tray? What items do you have in your possession right now that could be replaced with reusable alternatives (tissue, disposable pens, juiceboxes, etc.) Conduct a similar survey in your home.
   - Identify items in your household trash that can be recycled. What is the weight of these items? Brainstorm lists of ways you could decrease the number of items discarded by recycling.

Reduce:
Reducing solid wastes means decreasing the number and volume of discarded products. This includes reducing the amount of unnecessary packaging and reducing the amount of toxic chemicals in these products.

Reuse:
Precycling means that the consumer purchases long-lasting, reusable items, rather than disposables.

Recycle:
If properly treated, aluminum, batteries, corrugated cardboard, clothing, glass, paper and cardboard, plastics, oil, tin and steel cans, tires, and yard wastes (including Christmas trees) can be recycled.

Using figures from your team, estimate how much these recycled materials would be worth per year. According to the U.S. Environmental Protection Agency, the average prices in dollars per ton of some common recycled materials are as follows: cardboard boxes, 35; newspapers, 10; white paper, 45; aluminum cans, 1,050; steel, 55; glass, 40; clear-plastic soda bottles, 120.[2]

Average Price Per Ton of Recyclable Materials

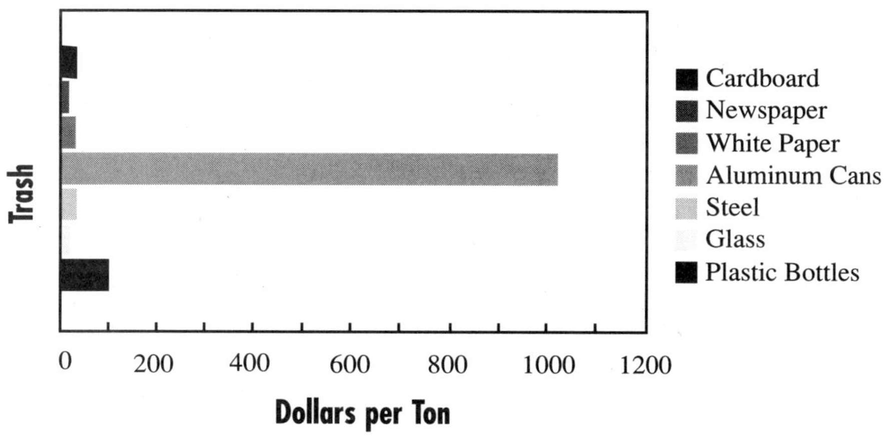

- Conduct a survey in a grocery store: Find examples of "overpackaged" products and their alternatives, or survey your homes for examples of environmentally friendly or unfriendly choices in packaging.
- With your team, brainstorm causes of the solid-waste problem. Use these ideas to help you design and conduct a consumer survey to assess consumers' perceptions about the cause of the solid-waste problem.

2 *1992 Information Please® Environmental Almanac,* World Resources Institute, pp. 123–124.

Optional Extensions

1. Conduct a class debate using the *USA Today* article below.

The Garbage Glut Is Everyone's Business

Source: *USA Today,* November 17, 1989[3]

Garbage just won't go away. And we can't ignore it anymore. There are fewer and fewer places to dump it.

The "pooh pooh choo choo"—5,000 tons of sewage sludge—is chugging its way back to Baltimore after a month-long trip, first to Donaldsonville, La., where residents turned up their noses, then to Moss Point, Miss., where the welcome mat was also yanked.

Five train cars of rotting refuse from Philadelphia headed for Brazil, Ind., met the same fate last month. A court order turned it around and sent it back home. And in Los Angeles, trash trucks hustle from one dump to another, trying to find one that hasn't taken in its daily quota. Unless something is done by 1992, the city will be turning out 6,000 tons of garbage a day with no place to put it.

Across the USA, city and county officials are longing for the good old days, when the mountains of garbage produced by conspicuous consumption were whisked away, never to be whiffed again. But garbage that used to be dumped, burned, then buried is now burying us. Landfills, where most trash winds up, are disappearing at an alarming rate. The Environmental Protection Agency expects that 2,000 of them will disappear within four years. What's worse, landfills have not proved to be the busy little biodegraders that people expected. They've actually preserved hot dogs and carrots for 15 years; 20-year-old newspapers can still be read.

What can be done? Plenty. We can produce less trash. More firms are finding ways to reduce packaging waste. And the federal government has set a goal of reducing garbage production by 25% by 1992. We can convert more trash to energy. Tazewell County, Ill., hopes to turn methane gas at its landfill into electricity for 2,400 homes. More communities need to seek ways to convert waste to power.

[3] "The garbage glut is everyone's business," *USA Today,* November 17, 1989. Copyright 1989, *USA Today.* Reprinted with permission.

And we can recycle more of our trash. Thirty states are implementing laws to separate trash for recycling; 10 have mandatory separation. The recycling idea is catching on. Two-thirds in a recent poll said they had recycled some garbage. Minnesota dairy farmers are finding that shredded newspapers make great cow bedding. School cafeterias and fast-food restaurants earn money by recycling plastic foam products. New Yorkers return three-quarters of their bottles and cans for deposits each year, keeping 100 million tons of trash out of landfills.

Such efforts show that individuals can make a huge difference. But the writer across the page thinks the price is too high. He'd rather pay someone else to do his sorting. The price may be steeper than he thinks. In Woodbury, N.J, he'd pay a fine of up to $500 a day for not sorting his trash. In Hamburg, N.Y., he would forfeit trash pickup. The price is steep because the consequences are dire: We can be buried in our own refuse if we don't act soon. Ignoring garbage won't make it go away. Dealing with it will.

It's a dirty job, but everybody has to do it.

2. Compare and contrast the results of your teams' studies with results from other schools in your global community.
3. Have a "trashless" picnic. Challenge each other to bring a lunch that produces no trash.
4. One person's trash is another person's treasure. Sponsor a competition for the most creative reuse of what would otherwise have been a "throwaway." Or give an award monthly for the most innovative suggestion for one of the three Rs.
5. If there is a curbside recycling program in your community, find out as much as you can about it (cost, volume, what is recycled, level of acceptance, etc.). Devise an advertising campaign to raise the level of participation in your community.

Explanation

ACTIVITY 5.3

How Long Will It Be There?

In this activity students design and conduct a long-term investigation of the rates of decomposition of various types of household solid waste under various conditions.

Objectives

- Determine the rates of decomposition in a landfill of various types of household solid waste
- Design and conduct an investigation that demonstrates knowledge of independent and dependent variables and controls

Materials

Examples of household garbage, Student Log page 217; materials for experiments will vary

Procedures

1. Hold up examples of items found in a typical household's garbage. Ask students to speculate about how long they think it would take for items to decompose. Record their predictions on a piece of newsprint and post them in the classroom.
2. Invite students to ask questions about how and why materials in a landfill decompose, and what factors affect the rate of decomposition.
3. Instruct each team to design an experiment to study the rate of decomposition of one or more types of household trash in a landfill and the factors that affect the rate of decomposition. Students may choose to actually bury items in the school yard (about 6 inches deep) and exhume them at later intervals or simulate "burial" in containers (Petri dishes or comparable containers) in the classroom. As each team designs its study, discuss controls, dependent and independent variables, and how they plan to report the results of their experiments to the rest of the class. Emphasize that this study will be conducted over several months, since some items may decay very slowly.
4. When the experiments have concluded, students should compare their experimental findings with their original predictions and report their findings to the class.
5. Combine the class data and make a time line for decomposition of various types of solid waste. Include milestones in students' own lives as a frame of reference. (For example, imagine a bag of trash buried on the day they were born; when the wood decomposes, they will have graduated from high school!)

6. Invite students to explore their feelings about solid waste. Have any students made any changes in their behavior as a result of this project?

Optional Extension

Start a project in the community to raise awareness and encourage people to practice the Three Rs. Brainstorm ways to teach what you've learned to members of your community and to encourage them to change their behavior. Each team selects one way to do this and follows through on the project. Report your progress on the Global Thinking Project bulletin board and ask teams in other schools to do as well.

ACTIVITY 5.4

Evaluating Waste-Disposal Options

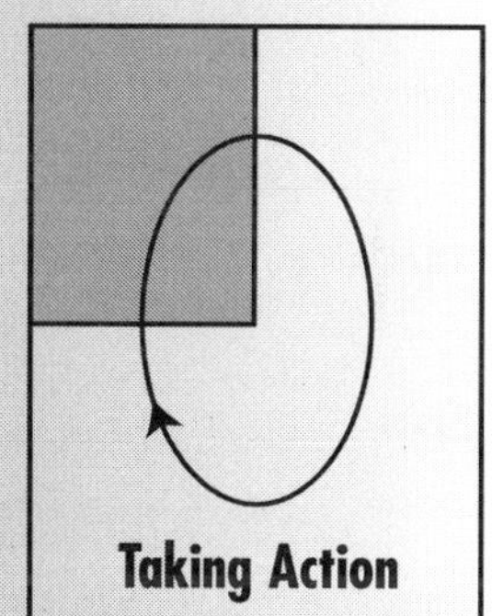

In this activity students develop evaluative criteria and evaluate the common options for disposing nontoxic solid waste.

Objectives

- Distinguish between environmental problems and issues
- Generate a list of criteria (based on values) that can be used to distinguish between alternative solid-waste disposal methods
- Evaluate solid-waste disposal methods based on evaluative criteria developed in class

Materials

chart paper, markers, access to data sources (library, Internet, etc.), Student Log pages 218, 219

Procedures

1. Begin by asking students whether they feel that solid-waste disposal is an environmental problem.
2. Ask why this problem has become an issue. (This will require that students distinguish between a problem and an issue, which results from disagreement about the solution to a problem.)
3. See whether the students can list the three common methods of disposing of nontoxic solid waste: open landfill, sanitary landfill, and incineration. Solicit knowledge of these.
4. Create three expert groups to investigate each of the three methods of solid-waste disposal. Before the groups begin their investigations, work with the class to develop a list of questions about solid-waste disposal methods that they'd like to have answered before evaluating

Environmental problem: a problem that is the result of the interactions between humanity and the environment

Environmental issue: an environmental problem about which there is disagreement as to the solution

Open landfill: a large area of land that is excavated and filled with waste

Sanitary landfill:
a landfill that is lined with a nonporous material to prevent liquids from leaching into the surrounding soil, and in which each day's deposits of waste are covered with a layer of dirt

Incineration:
disposal of wastes by burning

them. Post these questions on chart paper in the classroom. Invite students to add questions as they think of other things during the course of their investigations. Also, discuss possible sources of information for their research.

5. Allow time for each of the expert groups to present the results of their investigation to their teams.
6. Ask the students in teams to discuss the slogan "Not in my backyard" (NIMBY). What values do you think a person making that statement holds? What values would a person opposing that viewpoint hold?
7. Use the lists of values generated in the previous discussion to come up with a class list of values that could be used to evaluate solutions to the solid-waste disposal issue. Examples might be cost, health hazard to humans, convenience, and so on.

   Using the decision-making grid below, ask students in teams to evaluate open landfills, sanitary landfills, and incineration against their criteria.

	Criteria			
Alternatives				
• Open landfill				
• Sanitary landfill				
• Incineration				
• Source reduction				
• Recycling				
• Other				

Decision-Making Grid
Students should generate criteria in the grey cells. Examples might be cost, health hazard, convenience, and so on.

Each team should come up with what its members think is the best solution for their community.

Allow time for teams to present their decisions and rationales to the class.

Optional Extensions

1. Discuss opinions and conclusions with students in another school. Compare and contrast their reasoning with yours. Did any values enter into one class's decisions and not the other's?
2. Discuss with the class ways they could share their opinions with decision makers. Students may wish to write letters, invite a decision maker to class, stage a public debate, produce a videotape for a PTA meeting, and so on. Encourage individual teams to follow through on an action project.

3. Follow your trash—find out where your personal household trash goes when it leaves your house. (This may involve interviewing the people who pick up the trash in your neighborhood and asking a parent to drive you to the final destination of the trash.) Make a graph or other visual aid to communicate your findings to your classmates.
4. Gather articles from local newspapers dealing with the solid-waste disposal problem in your community. What are the issues? Compare with information gathered at other schools. Are the issues the same in all communities?
5. Visit an incinerator, landfill, or recycling center in your community.

CHAPTER 6

Project Water Watch

I have been in love with the sky since birth. And when I could fly, I wanted to go higher, to enter space and become a "man of the heights." During the eight days I spent in space, I realized that mankind needs height primarily to better know our long-suffering Earth, to see what cannot be seen up close. Not just to love her beauty, but also to ensure that we do not bring even the slightest harm to the natural world.

Pham Tuan
Astronaut, Vietnam

A satellite view of rivers

The activities in Project Water Watch are designed to show your students how to investigate and explore the realm of water on the Earth. The activities can focus on any available body of water, such as a pond, lake, creek, or river.

We suggest that you choose a body of water that is located close to your school for convenience. If there is a body of water on the school property, we recommend that you use it for study. (See the planning chart below to see how time for each activity should be allotted.)

Goals for Project Water Watch

1. Students will become familiar with their watershed area, the rivers and other bodies of water within it, and possible problem spots.
2. Students will become familiar with the major aquatic macroinvertebrates, their lifestyle, and their role in the river food web.
3. Students will conduct chemical and physical investigations of aquatic environments.
4. Students will suggest ways that river quality can be improved and will explore avenues through which they can take action.
5. Students will network with other schools that are part of their global community. Networking activities will include sending other schools data on the quality of their water test site, analyzing other schools' water test sites, and engaging in collaborative projects.

Planning Chart

Activity		Class Periods
6.1	Where does your river come from; where does it go?	1–2
6.2	Field trip: Chemistry of the river	2
6.3	Making sense of river data	1–2
6.4	Field trip: Aquatic macroinvertebrates	1–2
6.5	Macroinvertebrates as bioindicators of health	1–2
6.6	Protecting our river	1–3

Web Site Information

From GTP home page (www.gtp.org) click on "projects" to access the Project Water Watch home page.

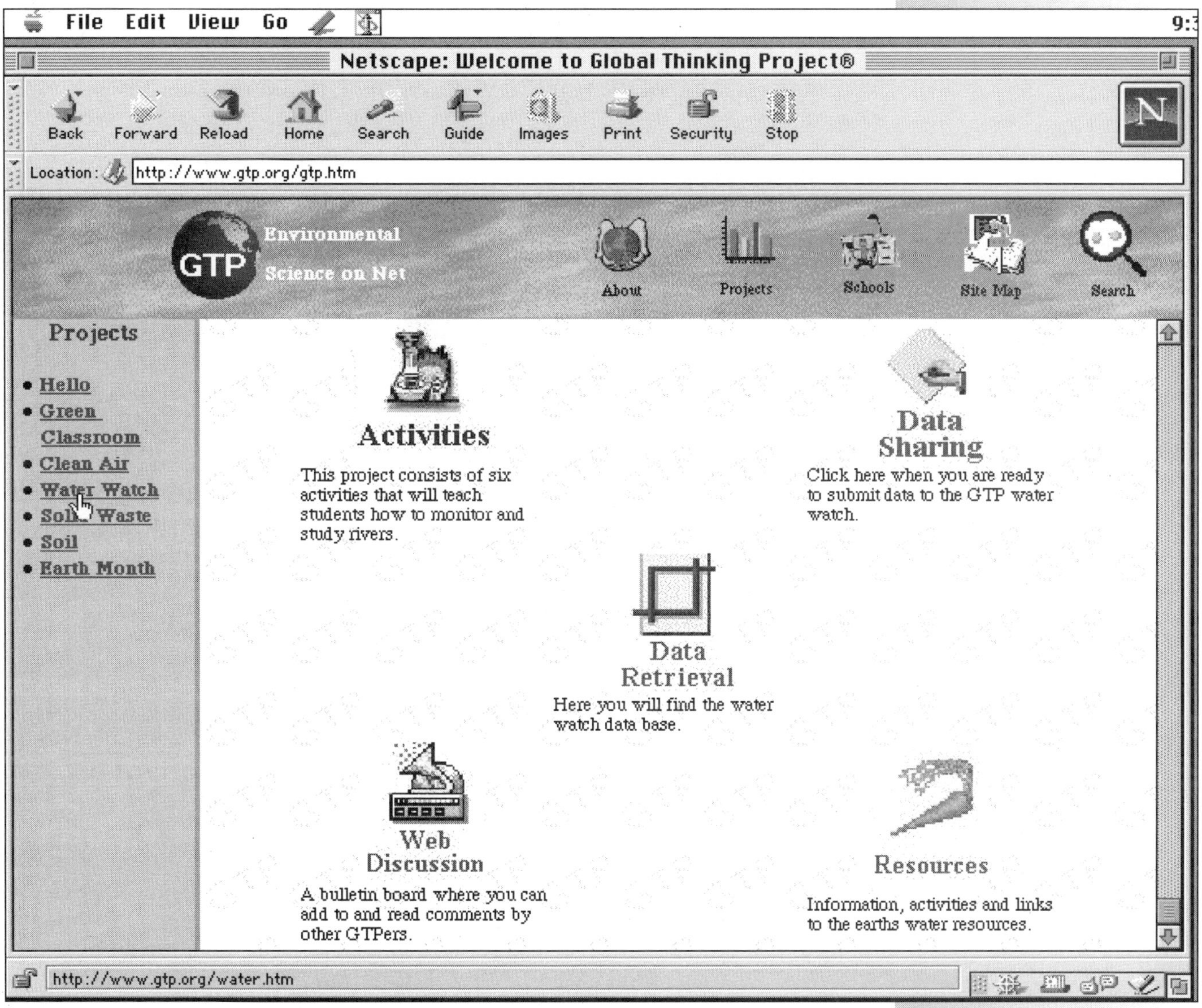

Project Water Watch Home Page

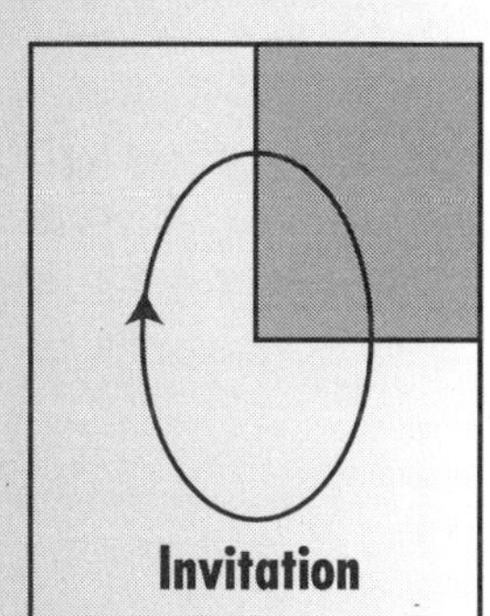

ACTIVITY 6.1

Where Does Your River Begin and End?

Monitoring a river is like taking the pulse of a continent. How healthy our waterways are can tell us much about the way we live. However, just as our bodies are very complex and difficult to understand, so are our rivers. The following activities only scratch the surface of river ecology, and one must be careful about drawing hasty conclusions.

The following activity is designed to allow students to identify the source of their water supply and, if possible, to trace this source back to its origin. By examining the source of their water supply, students will become aware of the physical link they have with other students living many miles away. In addition to the human connection, students will become aware of possible industrial or agricultural sources of water pollution.

Objectives

- Use maps to trace a local river, its tributaries, and its destination
- Locate cities, factories, and other potential pollution sources on the map
- Generate research questions and select study sites by using a map

Watershed:
a line that separates source streams that flow into different river systems. In the United States the term divide, as in "the Continental Divide," is synonymous with watershed.

Ground water:
water that fills pore spaces and the space between layers of rock beneath the surface of the earth

River:
a natural stream of water that follows a definite course, usually into an ocean

Reservoir:
a pond or body of water where water is stored for later use

Materials

maps showing local rivers, waterways, and their tributaries (one topographic map per class, one regional map per team); maps of local and regional watersheds from the Internet, information on cities, factories, farms, and other potential pollution sites (e.g., newspapers and chamber of commerce information), highlighters (two different colors per team); Student Log page 220

Procedures

1. Open the activity by holding up a glass of water and asking the students, "Where did this come from?" Encourage them to trace the water back from the tap to the actual supply source.
2. Briefly explain to the students that there are a variety of water supply sources. See if they can name them. (answers: ground water, rivers, and reservoirs)

   Pass out a map of the region to each team. Have students locate their city or town and its water supply. Then ask students to trace this supply source (or its tributaries) back to the origin. Students should use a highlighter to mark this route on their maps.
3. Having traced their water supply back to its origin, have teams identify and list
   - The major cities/towns upstream from their water supply source
   - Industries or rural farmlands located on the water supply route. Students should list contaminants that might come from these sources that might find their way into the water supply.
4. Pour the glass of water down the drain. Ask the students to trace its journey downstream. (Teams should mark the route in a different-colored highlighter on their maps.)
5. Ask teams to discuss the following questions and be prepared to report on their discussion to the class:

- What is the value of knowing where your drinking water has been and where it is going?
- In terms of water purity, where would you find the purest and cleanest drinking water, at the origin or at the end of a water-supply source? Why?
- What questions do you have about the quality of your local water supply, now that you know something about its source?

6. Ask cooperative groups of students to use their questions to suggest a study site and a plan that may provide answers to their questions.
7. Hold a class meeting to decide which question the class wants to try to answer. Practicality will be important. Easy and near access will be important. Consider also how many times a year a river should be tested. (When does nature change the "quality of the river"?). Rivers should be monitored four times per year, corresponding to each season. If you can arrange to visit the same river during two different seasons, students will be able to compare river-quality data for different seasons.

The Dirty Water Cleanup Challenge!

To culminate Activity 6.1, you might want to involve your class in the Dirty Water Cleanup Challenge. This is a hands-on laboratory activity in which students find the best method to clean up some "dirty water" that you provide. After students complete their research, they can send their results to the Project Water Watch home page.

In the Dirty Water Clean Up Challenge, students will find the best method(s) for cleaning a sample of dirty water.

Materials

small cups (half with holes in them), sand, cotton, charcoal chips (Before class, prepare a gallon of water that contains 20–25 drops of blue food coloring, 1/2 cup of soil, and 1/4 cup of vinegar, shaken.); Student Log pages 221, 222

Procedures

1. The owner of a local fish pet store wants to use some water from a pond in the store's aquariums. The water is very dirty. The owner needs your help to find a method that will take most of the things out of the dirty water. While reading this challenge, show the students the "dirty water." Tell the students that they can use the following materials when they perform their experiments: small cups with holes in them, sand, cotton, and charcoal chips. Show them how to make a simple filter. (In a cup with a hole in it, place successive layers of cotton, sand, and charcoal. Pour some of the dirty water through the "filter" into another cup.)

Water Filter System

In a cup with holes poked through the bottom, place layers of cotton, sand, and charcoal. Pour dirty water through the layers. The filter will "clean" the dirty water.

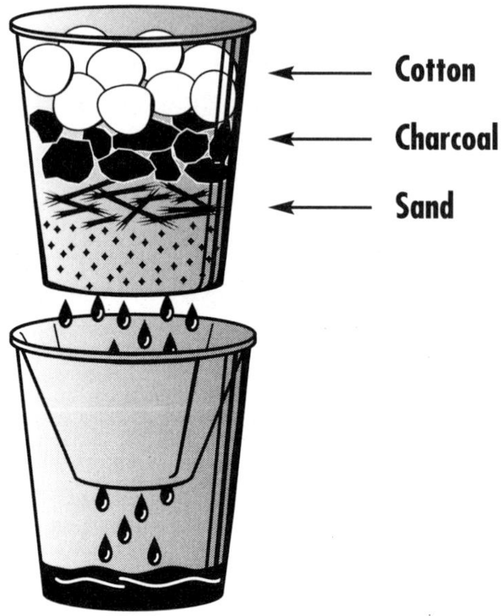

2. Tell the students to find a method to clean up the water. The students should work in small teams. They should:
   - Write a question that can be answered with the materials. The question could begin with "What will happen if I . . . ?"
   - Write predictions of what they think will happen.
   - Write how they are going to do their experiments (include drawings).
   - Keep a record of their data. They can make charts and drawings.
3. Students should prepare a brief report of their results. Have a class discussion in which each team presents its results.
4. Students should report their results by posting brief reports on the Project Water Watch home page. (From the home page, point to Web Discussion, and write your reports.) How do your students' solutions compare with those of students from other schools?

Dirty Water E-mail Sample

Dirty Water E-mail Report
To: Global Community
Re: Best methods to clean up dirty water

____

____

____

____

Signed

Optional Extensions

1. Have students visit the Project Water Watch home page and explore some of the resources on water that are linked to the page. You will also find a scavenger hunt, which you might challenge students to complete.
2. Conduct an investigation of the local water-treatment plant to understand how water is treated before it is allowed to be dispersed as drinking water.
3. Visit a sewage-treatment plant to learn how waste water is treated before it is released.
4. Conduct a survey of other GTP schools to determine the most common source of drinking water. How many students obtain their drinking water from ground-water wells? rivers? reservoirs? bottled water?

ACTIVITY 6.2

Field Trip No.1: Observing the Physical and Chemical Characteristics of a River

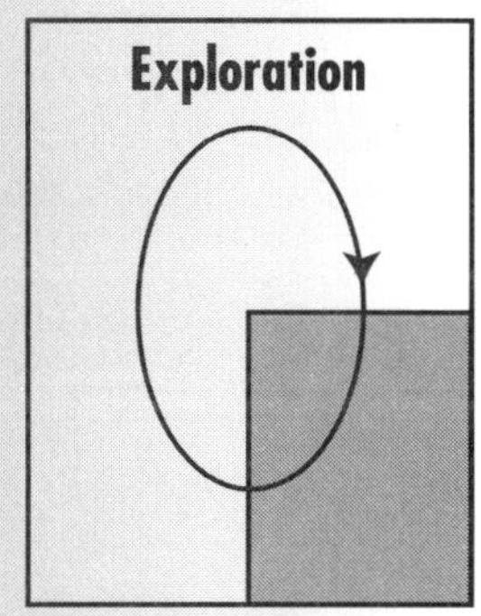

The physical, biological, and chemical qualities of a body of water such as a river are very complex and are constantly changing. To perform a "complete" analysis can be very expensive. In this activity students visit their river and describe its quality using quantitative and qualitative methods, with an emphasis on using their senses. Students develop a quality profile of their river and share their results over the Internet.

Objectives

- Visit a real river or another body of water chosen for study
- Learn a variety of methods for investigating a river
- Collect data on the physical and chemical characteristics of a river
- Discuss the significance of their water test results

Materials

Dissolved Oxygen Test Kit from CHEMetrics (See Resources for ordering information.); pH paper; meter stick, stop watch; thermometer; rubber boots and gloves, safety goggles; collection devices (jars, pails to bring samples back to the classroom); pencils, pens, crayons; camera, video tape machine (optional); Student Log pages 223–227

Advance Preparation

You will need to decide upon a body of water that your class will use as their test site. We recommend that you find a site close to your school. It does not have to be a large body of water. For safety considerations, we recommend a small stream that is readily accessible to you and your class.

Visit the test site ahead of time to determine the best way to access the stream and whether it can adequately handle all students you plan to bring. Ideally, the site should allow all teams to conduct their studies in close proximity.

Safety Considerations

1. Do not take students to bodies of water that have the least bit of risk in terms of increased water flow.
2. If students are going into the water, bring rubber boots and gloves for those who will enter the water.
3. If your sample site is below (downstream from) a waste-treatment site or the waters are known to be polluted, make sure all students wear rubber gloves and wash their hands after exposure.
4. Students should be organized into pairs or small teams. NO ONE SHOULD BE OBSERVING OR COLLECTING ALONE.
5. Be aware if any of the students do not know how to swim. Take appropriate precautions.

Procedures

1. Organize your class into work groups, each of which will be responsible for observing and collecting data on the physical and chemical characteristics of the river. Each team should work as an independent unit, collecting data on the section of the river assigned to it. In the next activity, students will analyze their data by using tables and graphs to make sense of their observations. The data from each group will be compiled, and the class will work together to synthesize findings about the site and send them to the GTP database on Project Water Watch. Go to the Project Water Watch home page to find the link for data sharing.
2. Before you go to the river, go over the observations and tests that each group will perform at the stream.

   Dissolved Oxygen (DO) Test: This test is a measure of the oxygen in solution in water. Each group will make two measurements of dissolved oxygen, using the CHEMetrics self-filling ampoules.The ampoule draws water into its chamber where the water mixes with an indicator that changes color. The oxygen level is determined using a colorimetric scale. Students will report dissolved oxygen in parts per million. Cold-water fish require DO levels of not less than 8 ppm during early life stages and not less than 4 ppm for other life stages; warm-water fish require DO levels of not less than 5 ppm for early life stages and not less than 3 ppm for other life stages.

pH Test: pH is a measure of the hydrogen ion concentration in water. It is measured on a scale from 0 (most acid) to 14 (most alkaline), with 7 considered neutral. Each group should make two readings and average the results. The pH will be measured using a short-range pH paper. Students need to be sure not to touch the end of the pH paper used to test the water. To protect freshwater fish life, the pH range should be 6.5–9.0.

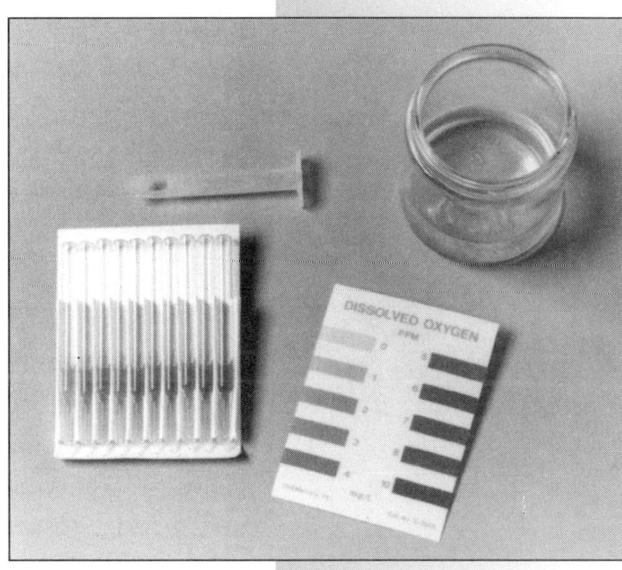

CHEMetrics® Dissolved Oxygen Ampoule
The ampoule is used to determine the amount of dissolved oxygen in the water. Students break the tip of the ampoule to allow a measured amount of water to enter the tube. Comparison to a color card reveals the amount of oxygen in ppm.

Water flow: Water flow is a measure of the velocity of water in a river. Water flow should be reported in meters/second. To measure the flow, have students mark off a 10–meter length along the river in their area. Use a cork (or a leaf) and watch to determine the time in seconds that the cork takes to flow 10 meters. Report the velocity as *m/sec.* To compute the velocity, sample the flow twice and take the average.

Depth: Use a meter stick to determine the depth of the water in the site area. If it is safe to do so, have the students determine the depth at various points across the river. Students can use this information to draw a cross section of the stream.

Temperature: Students should measure the temperature from two samples and take the average. Report in degrees Celsius. Since there will be a great variation in the thermometers that are used, you should have the students "calibrate" their thermometer. This is very easy to do and fun. Give the students a mixture of ice and water in a small cup. To calibrate the thermometer, have them stir the thermometer in the ice-and-water mixture. The temperature should read 0 degrees Celsius. If, for example, the temperature reading is +1˚C, then the students should make sure that when they report water temperatures for their river, they subtract 1 degree from the measured temperatures.

3. At the river, teams should make their observations and record them on the Water Watch Data Recording Form (See pages 225–227 in the Student Log.)

Dissolved Oxygen:
the amount of oxygen that is dissolved in water. It is measured in parts per million.

pH:
a value on a scale from 0 to 14 that shows how strongly acid or alkaline a solution is. A solution from 0 to 6 is acidic and from 8 to 14 is alkaline. Pure water is 7 (neutral).

Project Water Watch Observation Recording Form Sample, Part I

Team Name: ____________________ Team Members: ________________
Measurement Site (circle one): river, creek, pond, lake, ocean, other: ______
Date: __________________________ Latitude: ______________________
Weather: _______________________ Longitude: _____________________

Quantitative Measurements

Air Temperature:
Water Temp.: ______ #1 ______ #2 ______ Avg.
Depth: ______ #1 ______ #2 ______ #3 ______ Avg.
Flow: ______ #1 ______ #2 ______ Avg.
D. Oxygen: ______ #1 ______ #2 ______ Avg.
pH: ______ #1 ______ #2 ______ Avg.

Qualitative Measurements (Circle one word in each row.)

Land use	wooded	meadow	park	residential	commercial
Erosion	none	gullies			
Ground cover	grasses	shrubs	trees	soil	
Odor	rotten egg	musky	oily	sewage	none
Algae color	n/a	light green	dark green	brown	blue
Bed-bottom color	gray	orange/red	yellow	black	brown
Surface water	clear	oily	foamy	milky	muddy
Discharge	yes	no			

Notes: ____

Project Water Watch Observation Recording Form Sample, Part II

Water Watch Observations

Sketch a segment of the river (or body of water). Include as many details as possible.

Draw a cross section of the body of water. Include as many details as possible, including the locations of plants and rocks.

Optional Extensions

1. Have students bring in water samples taken from bodies of water near their homes. Samples should be collected in small, well-capped containers (such as baby food jars).
2. If more elaborate instruments or test kits are available, a more complete analysis could be performed.
3. Encourage a local water-testing lab to take an interest in your project. They could provide backup testing (to determine the accuracy of your methods) and possibly might conduct some sophisticated testing for you (such as testing for PCBs, pesticides, heavy metals, and so on.)

Explanation

ACTIVITY 6.3

Interpreting and Sharing the River Data

In this activity students will analyze their data by creating data tables and graphs using the Alice Data Tool. Each group will contribute its data to the class, and the class will work together to synthesize the data. The class will also share its results with other members of the GTP by sending the data to the GTP Water Watch database.

Objectives

- Organize quantitative and qualitative observations of a river
- Interpret and discuss students' findings about their river
- Share data by sending it over the Internet to the GTP water database
- Organize the data obtained from other schools into a master table (spreadsheet)
- Map data from the water-watch community using the Alice Map Tool

Materials

computer with Internet access; Project Water Watch observation forms with data; chart paper, marking pens; paper, pencil, rulers; Student Log pages 228, 229

Procedures

Before beginning activity, download Alice Software from *http://teaparty.terc.edu/tech/tech.html.* (See Resources.) Note that this can take 20 minutes or more. Refer to Chapter 2 for details on Alice.

1. Each team should enter all of their quantitative and qualitative data into a class data table. Have students record their data with marking pens on a large class chart that all can see. Leave empty rows of data so that you can add data from other schools to use for future comparisons.

Part A: Quantitative Information

	Temp.	Time	Date	Site	School	Lat.	Long.	Depth	Flow	D. O_2	pH
1.											
2.											
3.											
4.											

Part B: Qualitative Information

	Land Use	Erosion	Ground Cover	Odor	Algae Color	Bed-Bottom Color	Surface Water	Discharge
1.								
2.								
3.								
4.								

Class Chart of Water Observations Sample
Make a class chart on poster board or chart paper that can be preserved for future use and reference. Students can record their data here, making it visible to everyone. They can also post the data from other schools using information from the GTP Project Water Watch database.

2. Use the data-sharing template from "Submit Water Watch Data" on the Project Water Watch home page.

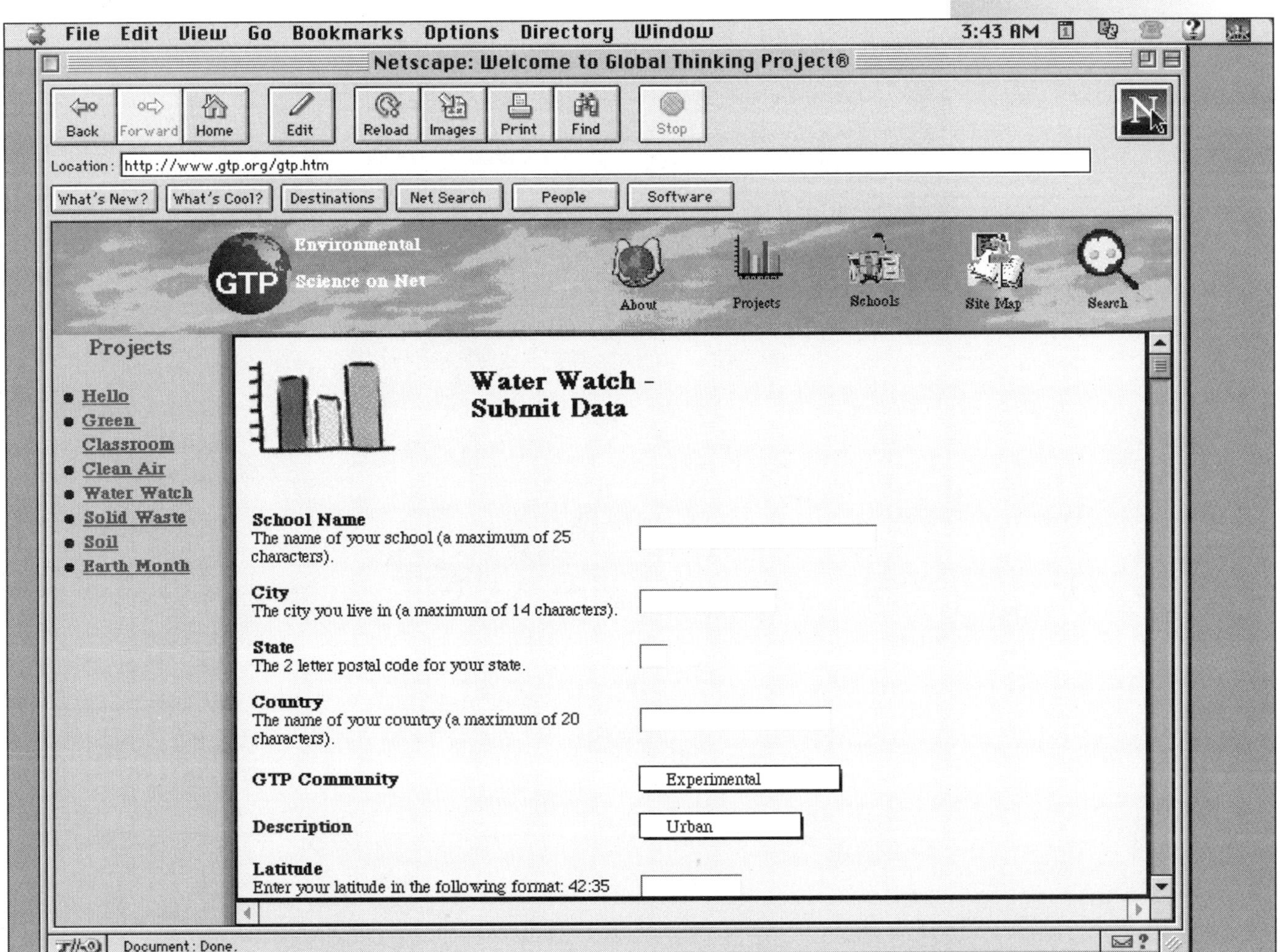

Water Watch Submit Data Form

3. As other schools contribute data to the Project Water Watch data base, retrieve the information from the GTP Web site. Go to the Project Water Watch home page, and click on "Retrieve Water Watch Data" from the home page. Download the data to make it visible on your computer screen.

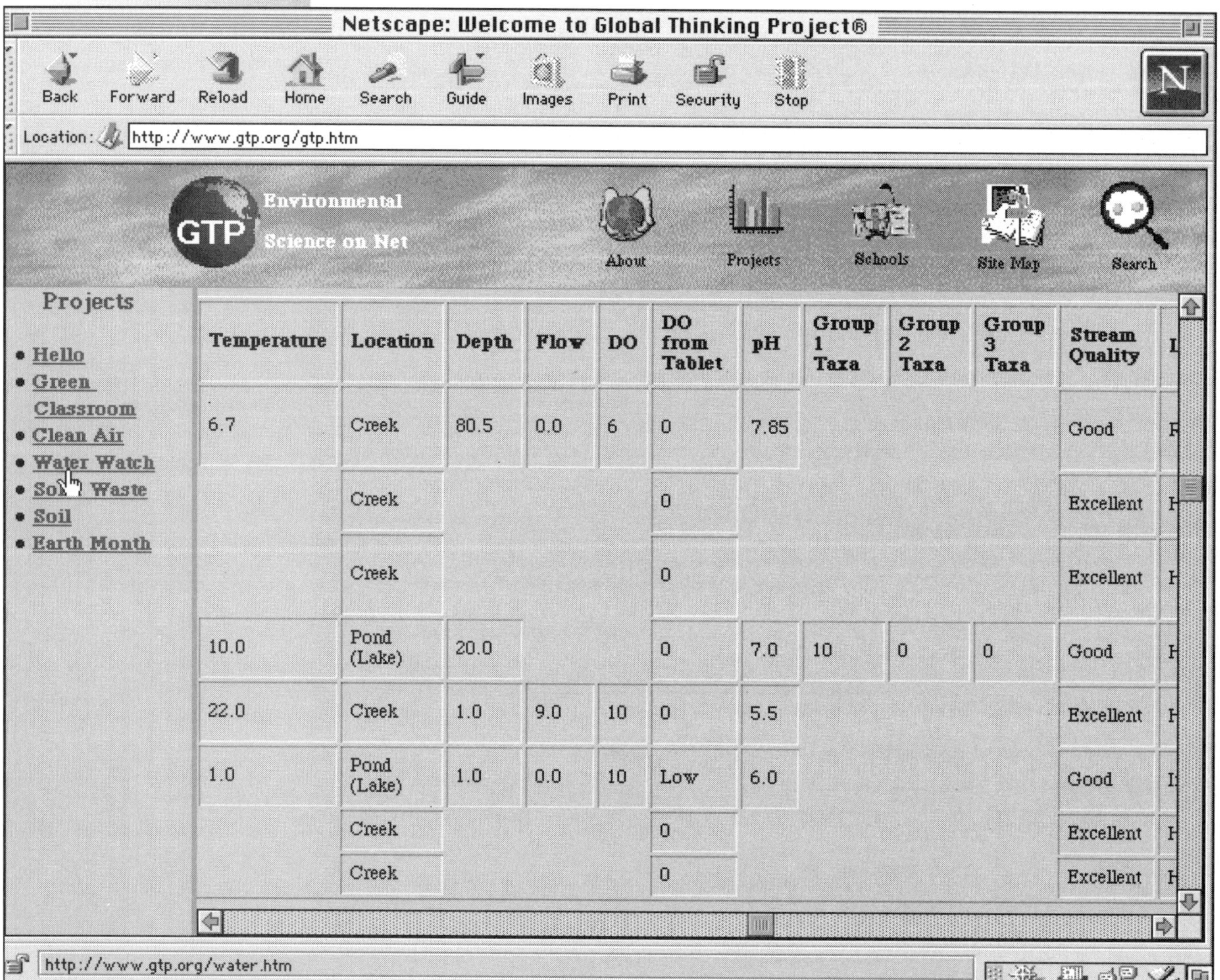

Temperature	Location	Depth	Flow	DO	DO from Tablet	pH	Group 1 Taxa	Group 2 Taxa	Group 3 Taxa	Stream Quality	L
6.7	Creek	80.5	0.0	6	0	7.85				Good	F
	Creek				0					Excellent	H
	Creek				0					Excellent	H
10.0	Pond (Lake)	20.0			0	7.0	10	0	0	Good	H
22.0	Creek	1.0	9.0	10	0	5.5				Excellent	H
1.0	Pond (Lake)	1.0	0.0	10	Low	6.0				Good	I
	Creek				0					Excellent	H
	Creek				0					Excellent	H

Water Watch Data Table

4. You will find reading the data convenient if you select some of the river data that was downloaded and transfer it to the class chart that you made earlier. This will make the data visible to everyone. Have students look at the data and develop a short list of questions for analysis. Questions can focus on comparison of measurements taken of rivers on different continents—differences in dissolved oxygen, pH, and temperature—as well as on some of the qualitative observations.

	Temp.	Time	Date	Site	School	Lat.	Long.	Depth	Flow	D. O_2	pH
1	12	9:00	2-2	creek	Salem JH	32 N	84 W	0.5 m	0.4	8	5.5
2	2	10:00	1-29	river	Mos 91	55 N	35 E	0.4	0.8	6	6.5
3	10	14:00	1-30	river	Lavinia	41 N	2 E	0.1	0.2	6	5.5
4	16	13:00	2-5	creek	Oberon	38 S	144 E	0.5	0.5	8	5.5

Project Water Watch Data Sample
Students should create a data table like this on a large sheet of paper, as well as use the Alice Data Retrieval tool. Qualitative observations should be added to the table to make it complete.

5. If you want to send a report to other schools via e-mail, go to the Project Water Watch home page and click on "Web Discussion." Students can write a brief report of:
   - The research they did on their river
   - Analyses made after studying the data submitted by other groups of students
   - Responses to other students' reports

Optional Extensions

1. Visit the GTP Web site and use the search tool to find and explore a river on another continent.
2. Suggest to the class that different groups analyze the data received from other members of the GTP. Students can make graphs of some of the data to answer questions about the physical and chemical characteristics of the sites studied. Use the graph to judge similarities and differences in the sites investigated.
3. Some students might like to use a large map of the world to post some of the interesting results obtained from other schools. They can identify the locations of the sites, and some of the salient characteristics can be reported upon.
4. Using the Alice Map Tool as a reference, students might begin a study of the location of important rivers in each of the Global Thinking countries. Students can construct a bulletin board showing the major rivers.

Macroinvertebrates: aquatic organisms lacking a backbone that can be seen with the naked eye.

ACTIVITY 6.4

Field Trip No. 2: Collecting and Identifying Aquatic Macroinvertebrates

Macroinvertebrates are "aquatic organisms that can be seen with the naked eye and lack a backbone."[1]

They include aquatic insects, as well as mollusks, arthropods, and annelids. In this activity students will collect and identify the macroinvertebrates present in the body of water they are studying. In Activity 6.5 they will learn to use this information to determine stream health.

Note: You might want to combine this field trip with the field trip to a river described in Activity 6.2 on page 109.

Objectives

- Collect, examine, and identify major groups of river macroinvertebrates
- Research the habitat and feeding requirements of the organisms that students collect
- Construct a simple food web as a simple model of their river ecosystem

Exploration

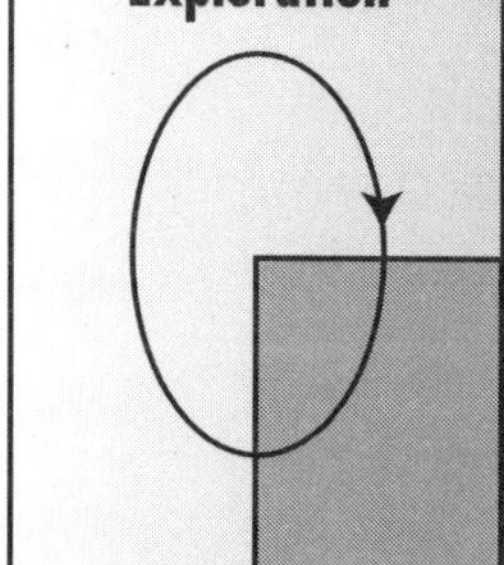

Materials

For each team:

white plastic dishpan; white soup bowls, compartmentalized polystyrene foam plates, or light-colored plastic trays (from some brands of microwave dinners); forceps; hand lenses; aquatic net (D-frame or triangular); empty gallon-sized milk jug (for filling with stream water); vials with screw caps or stoppers for storing specimens, rubbing alcohol; Student Log page 230

vial rack (can be made by drilling appropriate-size holes in a length of 2 × 4 lumber)

labels, pencil, data sheet (See page 120 for sample, and Student Log page 230.); clipboard

rubber boots, rubber gloves

laminated identification key (See page 121.)

first-aid kit

field guide

---

[1] McDonald, Brook S. and Mullins, Gary W. and Lewis, Stuart. "Macroinvertebrates as Indicators of Stream Health." *American Biology Teacher,* Vol. 53, no. 8, November/ December, 1991, pp. 462–466.

Procedures

Note: Select and visit the stream site in advance so you can be certain that it contains appropriate habitats for macroinvertebrates. Select a site that has a bottom composed mostly of small- and medium-sized rocks, cobbles, and gravel.

Caution students to avoid direct contact with the river water, in case it is polluted. Review previous safety reminders if necessary.

Each team should follow the procedures carefully. The class data can be pooled for Activity, 6.5 beginning on page 122.

1. Record the date, time, and site information, and draw a quick sketch of the collection site. This information should be recorded on the form shown in the Aquatic Macroinvertebrates Monitoring Survey.
2. Approach the collection site from downstream. As several students hold nets, the others should walk in front, picking up rocks and rubbing them underwater to dislodge organisms. After all rocks fist-sized or larger have been scrubbed, students should disturb the stream bottom with their feet, to allow collection of any organisms that have burrowed into the ground. When the collecting has been completed, the nets should be carefully scooped out of the water and carried to a clean, flat area for examination.
3. Empty the nets one at a time into a large white dishpan, and rinse with a small amount of stream water. As organisms emerge, carefully separate them into a small container (a bowl or small pan) that has a half inch or so of clear water in it.

   At this point, organisms can be taken back to the classroom for identification (they will keep for about 24 hours if refrigerated), or identification can be made at the site.

Aquatic Macroinvertebrates Monitoring Survey

Group: ______________________ Group Members: ______________________

Location: river: ___ creek: ___ pond: ___ lake: ___ ocean: ___

Date: ______________________ Weather: ______________________

Air Temperature: ______________________ Water Temperature: ______________________

Depth of Water in Collection Site: ______________________

Sketch of Collection Site:

Group One Taxa: Pollution-Sensitive Organisms; signify good-quality water		Group Two Taxa: Somewhat Pollution-Tolerant Organisms		Group Three Taxa: Pollution-Tolerant Organisms; can be in poor-quality water	
Organism	**Count**	**Organism**	**Count**	**Organism**	**Count**
Stonefly nymph		Crayfish		Aquatic worms	
Caddisfly larva		Sowbug		Midge larva	
Water penny beetle larvae		Scud		Blackfly larva	
Riffle beetle				Leech	
Mayfly nymph				Pouch snail	
Gilled snail		Danselfly nymph		Other snails	
Dobsonfly larva		Crane fly larva Dragonfly nymph Clam			
Total Group 1		Total Group 2		Total Group 3	

4. Sort the organisms into similar-looking groups, and identify as many as possible, using the Macroinvertebrate identification key[2] below and whatever field guides are available. (A good field guide in *Aquatic Insects* by Dennis Lehmkuhl, published by William C. Brown, 1979.) Enter the number of each type of organism on your data table.

Macroinvertebrate Identification Key

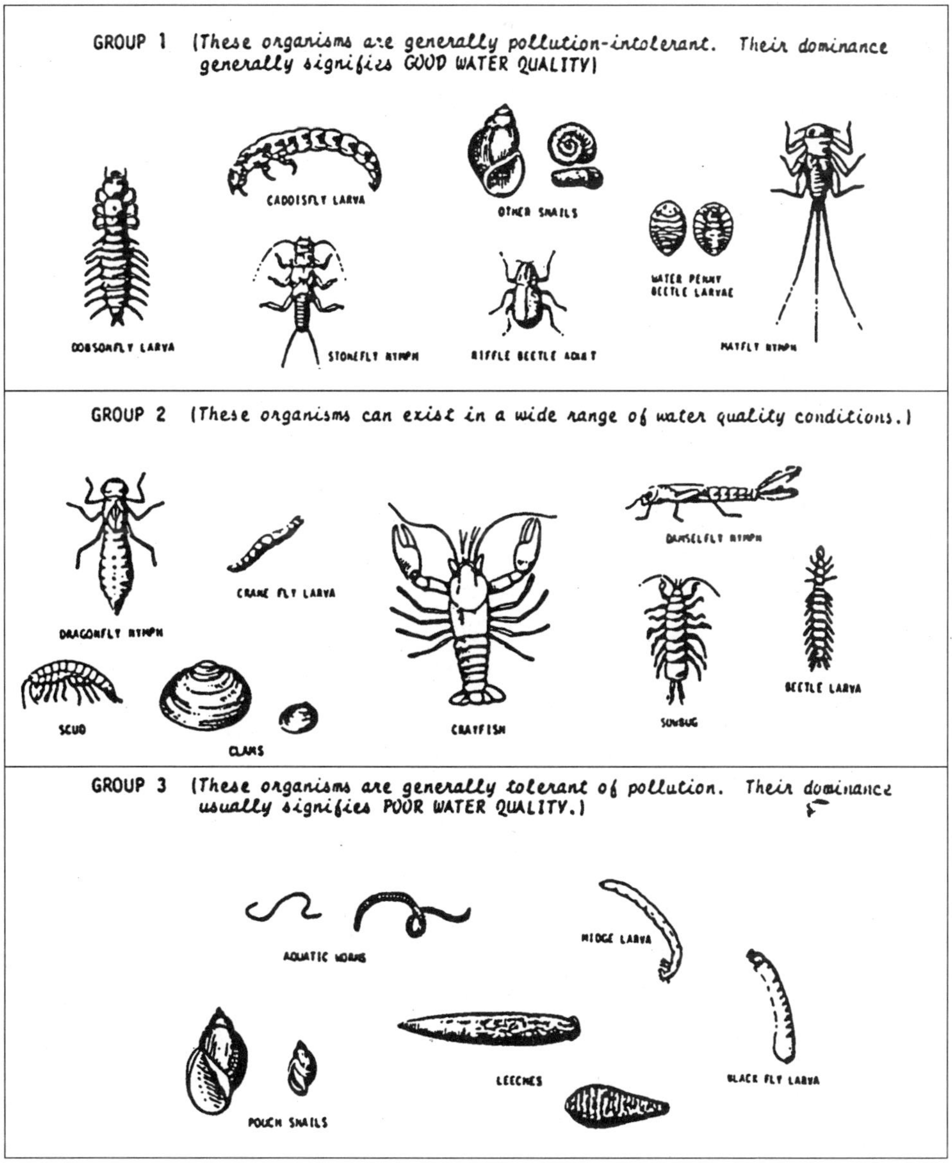

[2] Ohio Department of Natural Resources. Used with permission.

5. Transfer representative examples of organisms to vials containing isopropyl alcohol to start a reference collection. Label the vials with the date and collection site, as well as the name of the organism. Change the alcohol after two days, and check monthly thereafter.

Optional Extensions

1. Students may enjoy drawing detailed colored pictures of the organisms for display in their classroom.
2. Have students research food webs and then attempt to construct their own river food web. Students should be encouraged to include both organisms they collected and organisms they know live in the river but were not collected, such as fish and plants.
3. Construct a bulletin-board model of the river food web.
4. Prepare a dramatic presentation of the river food web.
5. Invite a fisheries biologist to give a talk on the fish and other life in the river.

Explanation

ACTIVITY 6.5

Using Macroinvertebrates as Bioindicators

In addition to physical and chemical parameters, the diversity of plant and animal life present is another indicator of the relative health of a body of water. Decomposition of organic pollutants such as sewage, effluents from food processing plants and paper mills, and agricultural runoff results in consumption of dissolved oxygen from a stream or lake. As dissolved oxygen is depleted, macroinvertebrates and fish that have high dissolved-oxygen requirements disappear, leaving behind a higher proportion of more pollution-tolerant species. Unpolluted bodies of water have a greater diversity of species present, while polluted ones have a less-diverse population, restricted to those species with relatively low dissolved-oxygen requirements. Thus, the relative numbers of pollution-sensitive and pollution-tolerant macroinvertebrate species present in a body of water can be used as a biological indicator of organic pollution.

Methods of assessing species diversity range from very simple to very complex. A simple analytical procedure is suggested here, with a more complicated one presented as an optional extension. In all cases, it is important to remind students that factors other than organic pollution may affect species diversity, and that it is important to gather many samples from a site at different times of the year before drawing firm conclusions.

Objectives

- Use observations of aquatic macroinvertebrates to assess stream quality

Materials

data collected in Activity 6.4, copies of the Stream Quality Assessment Chart, Student Log pages 230–233

Procedures

1. Count the number of types of organisms collected in Activity 6.4, and multiply the number in each group by the group index value.
2. Add the group index values together to find a cumulative index value for the stream. Use the Stream Quality Assessment Chart on page 124 to determine the quality of the stream.
3. After discussing the findings and using the results to estimate the quality of the stream you studied, go to the GTP Project Water Watch home page and send a brief report of your results to other GTPers. You might use the sample e-mail report as a model for your students' reports.

E-mail Report for Macroinvertebrate Data Sample

To: Global Community Schools
Re: Macroinvertebrate Report

Here are the results of our river study of macroinvertebrates.

<date>, <location>, <weather>, <air temperature>, <water temperature>, <depth>, <number of group one taxa organisms>, <number of group two taxa>, <number of group three taxa>, <latitude>, <longitude>, <school name>, <town>

We also found that: ______________________________

____

____

Signed

Optional Extension

Have students identify members of functional feeding groups (shredders, collectors, scrapers, predators) and determine the percentage of each present in your sample. Compare with data from other sites. (For a simple key, see *Field Procedures for Analysis of Functional Feeding Groups of Stream Macroinvertebrates* by K. W. Cummins and M. A. Wlizbach, Appalachian Environmental Laboratory Contribution No. 1611, University of Maryland, 1985.)

Stream Quality Assessment

The organisms in Figure 1 are grouped into three categories:

GROUP 1 (pollution-intolerant or good quality indicators)
GROUP 2 (organisms that can exist in both extremes of quality)
GROUP 3 (pollution-tolerant or poor quality indicators)

The organisms in these three groups are assigned a group index value.

GROUP 1 = 3 GROUP 2 = 2 GROUP 3 = 1

The analysis procedure consists of counting the number of types of organisms in each category and multiplying the group index value.

EXAMPLES:	**GROUP 1 TAXA**	**GROUP 2 TAXA**	**GROUP 3 TAXA**
	CADDISFLY(S)	DRAGONFLY(S)	BLACKFLY(S)
	STONEFLY(S)	CRAYFISH	MIDGE (S)
	MAYFLY(S) 3x3=9	CLAM(S) 3x2=6	2x1=2

The respective group index values are then added together to find the cumulative index value (which in the above case would be 17). By referring to the following chart, the stream quality assessment can thus be determined.

STREAM QUALITY ASSESSMENT	**CUMULATIVE INDEX VALUE**
EXCELLENT	23 and above
GOOD	17 - 22
FAIR	11 - 16
POOR	10 or less (SEE ACTION PROCEDURE ON PAGE 8)

NOTE: The organisms on the stream quality assessment form used in the field are to be recorded by placing a letter code in the corresponding block. Each letter represents your estimated count.

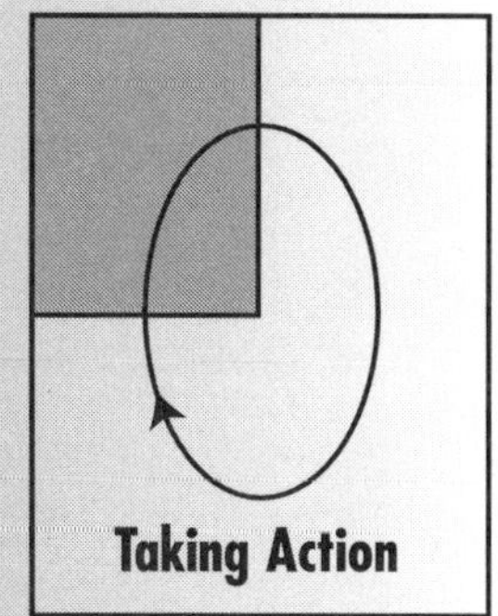

ACTIVITY 6.6

Protecting Our Rivers

A central goal of the Global Thinking Project is to provide students with opportunities to take action on their environmental concerns. The availability of water for human activity has reached a critical point, and students might find that even their own local water supply is threatened. It is not uncommon for municipal officials to tell citizens that they must boil their water before drinking it. In many communities old water-treatment plants are unable to manage the reclamation of water for citizens' use.

We've included an actual example of how a group of GTP students and their teachers not only studied a major river system, but made recommendations to city officials regarding the present water-treatment problems facing the city of Atlanta, Georgia, and its major river, the Chattahoochee.

The case is presented as an example of how you might work with your students to study a local river, especially if the river needs protection. In this activity we provide an example and steps that students can follow to act on their convictions and research. We call this kind of action *citizen scientist action.* We think you can use this example as a model for working with your students as they try to protect rivers in their region.

Objectives

- Discuss the actions taken by fellow GTPers to clean up the pollution of the Chattahoochee River, the major river running through Atlanta, Georgia
- Research water-quality laws that pertain to their river
- Learn what enforcement exists for river protection
- Find names of public and private organizations concerned with water quality and river protection
- Brainstorm actions that citizens can take to protect and improve their waterways
- Educate the public about the river ecosystem

Materials

phone books, newspapers, books on environmental organizations; paper, pencils, envelopes, stamps; poster paper, art materials; Student Log pages 234, 235

Web Site Information

In conjunction with the case study described here, you might want to visit the Chattahoochee River Project home page, which has a link from the Project Water Watch home page.

Procedures

1. In cooperative groups students should brainstorm a list of laws they think should be made to protect rivers. Have them discuss what should happen when someone breaks one of these laws. Then have students research the laws that exist and contact government agencies charged with protecting water quality to find out what laws and enforcement exist. Are their laws included in the actual law? Discuss whether the real laws are effective and what changes or additions seem needed.
2. Have cooperative groups investigate ways citizens can act to improve river-water quality (lobbying legislatures to change laws, demonstrating, forming citizen groups, cleanup actions, using the media, public education, lawsuits, talking to business leaders, etc.). What organizations are working in these directions?
3. Have cooperative groups decide on some action they could take to educate the public about their river. Educational posters or pamphlets on river quality and citizen action would be a good choice.

Optional Extensions

1. Invite representatives of public and private organizations (including businesses along the river) to talk to the group on river protection and ecology.
2. Conduct further studies in other locations and at other times of the year.
3. Develop a presentation (slides, a dramatic presentation of the river food web from Activity 6.2, etc.) to give at an elementary school or an organization.
4. Match up students with elementary students for a river-awareness field trip.

A Case Study: Protecting Atlanta's Chattahoochee River—Citizen Scientists in Action

Introduction

Atlanta's Chattahoochee River

In the Spring of 1997, fifty Russian students and fifteen Russian teachers came to Atlanta, Georgia, and were hosted by five schools to participate in the second half of a GTP exchange program. The Russian students lived in the homes of students whom they had hosted in Russia in March. Together, with the help of their teachers, they participated in monitoring activities at the school site, experienced a variety of school and cultural activities, and lived for twenty-one days in Atlanta. In order to bring all the students together on a common problem, the GTP staff proposed that each team of American and Russian students work on the Chattahoochee River problem. Each school pair would be assigned a piece of the problem in the form of one or more inquiry questions. They would use the resources of the Internet, make visits to the Chattahoochee, and interview experts to try to get answers to their questions. After ten days of research, the GTP organized a live conference using Georgia's GSAMS technology (Georgia Statewide Academic and Medical System). At three sites in Georgia, all one-hundred students and about thirty American and Russian teachers participated in a two-hour live television conference, during which they presented the results of their research on the Chattahoochee River. During the teleconference, we assigned the students their final task, which was to develop a set of recommendations from the standpoint of a different constituent group (homeowner, land developer, environmentalist) and present them at the GTP Summit, which was held at the end of the Russian visit. The details of these three phases, including the recommendations (presented in the form of a press release) are presented on pages 129 and 130.

[1] In research that we have conducted on this specific activity, students reported that working on a problem such as the Chattahoochee River issue led to a deeper understanding of the role of the citizen scientist in today's world. In particular, students felt empowered in that they could work toward a solution of a real problem, and thus they gained a positive outlook toward being a citizen scientist.

Phases

The *first phase* took place on May 7, 1997, at which time the ten participating schools in the GTP/Georgia/Russia Exchange engaged in a GSAMS teleconference. Each team presented the results of its research on one of the inquiry questions posed below.

The *second phase* occurred at the Environmental Summit at Simpsonwood on May 12, 1997. Using the research results of all the schools that reported at the GSAMS conference, each school pair prepared a set of recommendations to answer the question, "How can Atlanta improve its waste-water operation?" Each team was assigned a particular point of view (e.g., homeowner, land developer, environmentalist). Posters were prepared and students made presentations in a symposium format.

The *third phase* involved the presentation of the recommendations to appropriate officials in the city and state.

School Pair	Research Questions Assigned
Chamblee High School/ St. Petersburg School 157	What has been the role of state and federal regulatory agencies in the Atlanta sewer problem? What have they done to help resolve the sewer problem in Atlanta?
Chattanooga Valley MiddleSchool/ Chelyabinsk Lycee 93	What is the problem with the current Atlanta sewer system? What are the pollutants? the causes?
Dunwoody High School/ Moscow Gymnasium 710	How does the water leaving the Atlanta water works operation compare to the water that enters the Atlanta system? How is the water "cleaned"?
Lafayette Middle School/Pushino Gymnasium 2	What solutions have been proposed, and what has happened as a result of them? What has the city of Atlanta proposed as a solution to the problem?
Salem High School/ Yaroslavl Municipal School 1	Are the Atlanta water-treatment plants up to standard? If not, what is being done to make improvements?

Sample Press Release

Student Environmental Summit Yields Recommendations for Solving the Atlanta Sewer-System Problem

For Immediate Release
Tuesday, May 13, 1997

Contact: Dr. Jack Hassard
Georgia State University

jhassard@gsu.edu
404/651-2518

Atlanta, Georgia, May 13, 1997

Russian and American students participating in the Environmental Summit sponsored by the Global Thinking Project produced the following recommendations for solving Atlanta's sewer-system problem. The one-hundred students, representing five Georgia schools and five Russian schools, have been studying this problem for the past three weeks as a component of a three-week exchange project under the auspices of the GTP and supported by a grant from the United States Information Agency.

Recommendations by Lafayette Middle School and Pushino Gymnasium 2, from the point of view of homeowners:

1. Make strict regulations, more frequent testing, and daily checks to make sure water can flow freely within the system to prevent overflow.
2. Provide each homeowner with information about the status of street runoff and level of ground water in his or her area on a regular basis.
3. Contact the metro-area government to encourage replacement of the inadequate sewer system.

Recommendations by Chattanooga Valley Middle School and Chelyabinsk Lycee 93, from the point of view of land developers:

1. Build recycling plants to transform the solid human waste into fertilizer for farms.
2. Build another sewage-treatment plant to help in the areas that are becoming heavily polluted.
3. Make new standards for the size of the sewage lines that are to be newly installed or replaced in the city.

Recommendations by Chamblee High School and St. Petersburg School 157 from the point of view of elected officials:

1. Fine violators of pollution standards. Fines should be proportional to the amount of pollution.
2. Create an organization that will:
   - Research the pollution levels in the river
   - Propose laws concerning the Chattahoochee River and the Atlanta sewer system
   - Familiarize itself with the proposals of other organizations concerning the environment
   - Veto all other projects that may bring harm to the environment
   - Provide funding for approved proposals
   - Have the right to allow its lawyers to prosecute all offenses concerning Atlanta's sewer system and the Chattahoochee River Projects for the organization
   - Creation of storage bins for excess rain water
   - Renovation of old treatment plants
   - Funding of a research team that will explore new technology that removes E. coli from the water
3. Create a superfund supervised by the above-mentioned organization that will finance both the project accepted by the organization and the running of the organization. Financial support of the organization should be provided by:
   - EPA and other environmental organizations
   - Private contributions
   - Fines and environmental taxes
   - Money from plants and factories
   - Government-allotted funds
4. Cooperate with the mass media to obtain full coverage of the organization's actions toward solving environmental problems
5. Replace old portions of sewage pipes

END OF PRESS RELEASE

CHAPTER 7

Project Soil

Humans really don't think what is under their legs. They are always moving . . . And the best thing they do—they look how beautiful the clouds are above their heads . . . And they never look down, never praise . . . how beautiful the soil is!
Karel Chapek
Russian Poet

Bryce Canyon

Fertile soil has always been a feature of successful civilizations. In addition to producing food, fertile soil is necessary for the production of wood, natural fibers, and other components of consumer products. Yet in modern times we rarely stop to consider what is in the soil, where it comes from, and whether and how we must protect it.

Weathered bedrock is the parent material for soils. Over time, rock is weathered and mixes with organic materials and their breakdown products. In 1941 Hans Jenny identified five soil-forming factors—climate, organisms, topography, parent material, and time—that interact to produce the unique characteristics of different soils.

Soil is organized in layers, called horizons, that are approximately parallel to the Earth's surface. The horizons may be distinguished by properties such as texture, structure, consistence, and color.

Soil:
a mixture of weathered rock and decaying organic matter that covers the surface of the Earth and serves as a growth medium for plants

Planning Alert
For Activity 7.1, you will need to procure small soil samples from a variety of locations.

For Activity 7.2, you will have to have identified in advance an appropriate location for obtaining a variety of different types of soil samples.

The activities in Project Soil are designed to show students how to define and assess soil properties before creating and implementing a soil conservation project. (See the planning chart below to see how time for each activity should be allotted.)

Goals for Project Soil

1. Students will define soil.
2. Students will assess soil properties.
3. Students will plan and carry out a soil conservation project.

Planning Chart

Activity	Class Periods
7.1 What is soil?	1
7.2 Field examination of soil	1–2
7.3 Laboratory investigation of soil	1–2
7.4 Preserving the soil	1–5

Web Site Information

From GTP home page (www.gtp.org) click on "projects" to access Project Soil home page.

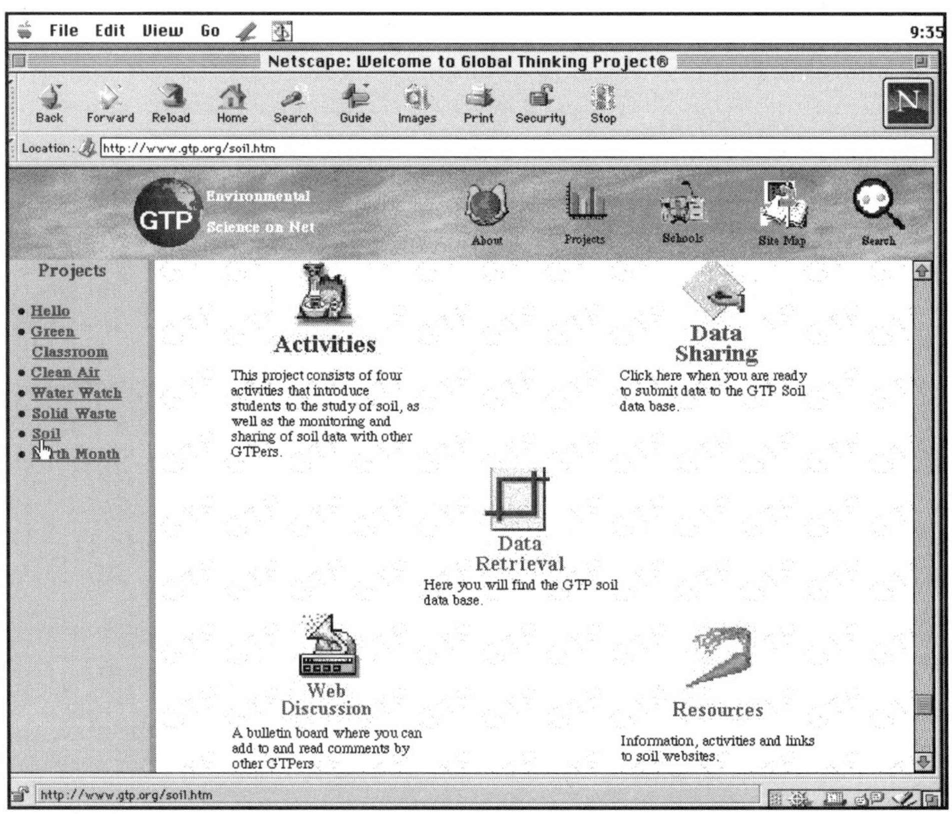

Project Soil Home Page

ACTIVITY 7.1

What Is Soil?

Students begin their investigation of *soil* by brainstorming what they know about soil, and where it comes from. The teams' lists are used to compile a class K-W-L Chart (what we Know, what we Want to find out, what we Learned), which is posted for the remainder of the unit. Students are introduced to the *five soil-forming factors* identified by Hans Jenny in 1941.

Objectives

- Generate questions about soil.
- Discuss sources of soil.
- Identify important factors in soil formation.

Materials

chart or construction paper, crayon or marking pens, masking tape; variety of soil samples, hand magnifiers; Student Log pages 236, 237

Procedures

1. Divide your class randomly into teams of four students for the duration of this project. Give each team a soil sample and give the teams five minutes to examine and describe their soil sample. Follow up by asking one member from each team to report their team's findings to the class.
2. Ask the students what soil is and why it is important.
3. Ask each team to complete a T-chart drawn on a large sheet of newsprint by brainstorming answers to the following questions:
   - What is soil?
   - What questions do you have about soil?

Soil T-chart Sample

What is soil?	What questions do you have about soil?

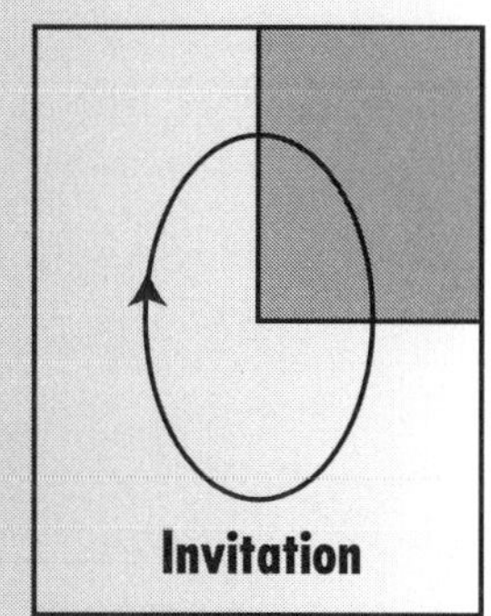

Soil Properties: soil characteristics that distinguish one soil from another. These may include color, texture, consistence, and structure.

Soil Horizons: distinct layers of soil with different properties that are parallel to the Earth's surface. A formal classification for soil horizons, which distinguishes primarily organic from primarily inorganic layers, may be found in soil-science texts.

Steps in Constructing a K-W-L Chart:

1. Help students brainstorm what they know or think they know about a topic.
2. List questions students hope to have answered as they work.
3. List what students have learned after investigation.

Five soil-forming factors:

1. Parent material: The type of rock from which soil is formed affects its chemical composition and texture.
2. Climate: Important climatic influences on soil formation are precipitation and temperature.
3. Organisms: Plants and animals affect soil formation as decomposers and consumers of organic matter, and as earth movers.
4. Topography: The degree of slope of a site may influence the quantity of precipitation absorbed, the rate of soil removal by erosion, and the movement of suspended and dissolved materials from one area to another.
5. Time: Weathering of rock and the subsequent modification of soils by organic processes takes place over relatively long periods of time.

Soil Survey: a systematic description, classification, and mapping of soils in a particular area.

Give the students ten minutes to generate a list of answers to these questions. When the teams are finished, tell them to look over their lists and select and circle the most important item on each list. Have the communicator from each team record these items on a class master chart. Use the responses to initiate a discussion with the students about soil and soil formation. Post team charts in the classroom for future reference.

4. Introduce the five soil-forming factors from the sidebar on this page. Many important properties of soil are determined by the parent material (rock) from which it was formed. What is the bedrock in your area? Discuss with the students their ideas about how the other factors (climate, topography, plants and animals, and time) might affect different soils formed from the same parent material. Examine and discuss soil surveying with students.

ACTIVITY 7.2

Field Examinations of Soil

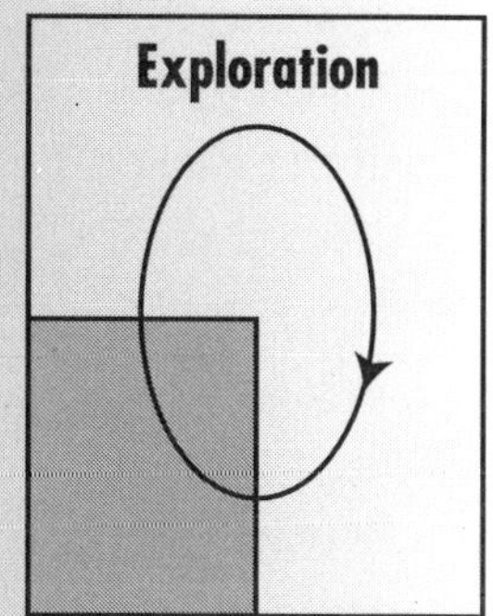

Objectives

- Learn methods of gathering soil samples systematically.
- Learn methods for field characterization of soil samples.
- Compare and contrast soil samples taken from different microclimates at a given site.

Materials

paper, pencils, clipboards, tape measure; compass, hand lenses, camera (optional); spades, centimeter rulers, self-closing plastic bags, marker pens or adhesive-back labels; standard soil-color chart, copies of the data table and soil structures diagram; wash bottle filled with distilled water; Student Log pages 238, 239

Advance Preparation

Locate a site from which you can collect soil samples. The site should be large enough for teams to be separated by at least 15 meters, and should provide a variety of microclimates for study.

Procedures

After each team selects its study site, it should complete the following steps:

1. Sketch and describe the site. Use a compass to locate North on your sketch. Include a description of the plants and animals you observe in the general area around your site. Use a hand lens to look for small insects and other organisms on the surface of the soil. Note if your site is on level ground, at the top or bottom of an incline, or on a slope. You might also note the approximate distance of your site from buildings, roads, ball fields, and so on. If a camera is available, photograph the site.
2. Dig a small pit several inches deep. Keep one side of the pit smooth so that you can see it in cross section. Observe the soil profile carefully and make a sketch showing the relative thickness of the layers. Note the distance from the surface and the depth of each layer in centimeters, and the color of each layer. Take a sample of approximately 150 grams from each layer and put it in a labeled self-closing plastic bag for later use in Activity 7.3.

3. Complete the data table on page 239 of the Student Log for each layer, identifying the top and bottom boundaries (in centimeters from the surface) and soil characteristics. (See explanations on pages 137 and 138.)

Data Table Sample

Team Name ______________________ Date __________

Brief description of the site: ______________________

____

____

____

Sketch of the site:

Sketch of pit, with horizons shown:

	Horizon No. 1	Horizon No. 2	Horizon No. 3	Horizon No. 4
Top boundary (cm from surface)				
Bottom boundary (cm from surface)				
Color				
Structure				
Consistency				
Texture				

- *Soil color* is influenced by the inorganic and organic composition of the soil. Determine the dominant color of your sample by referring to a set of standard color chips.
- *Soil structure* refers to the arrangement of the soil particles. Soil may be described as granular, blocky, platy, columnar, or prismatic, depending on the shape taken by soil aggregates (clumps), called peds. Soil is structureless if it is single grained (like beach sand) or massive (sticking together in a structureless glob). Examine a sample of undisturbed soil in your hand. Refer to the soil structure diagram[1] below to help you determine the structure of your soil.

Soil Structures

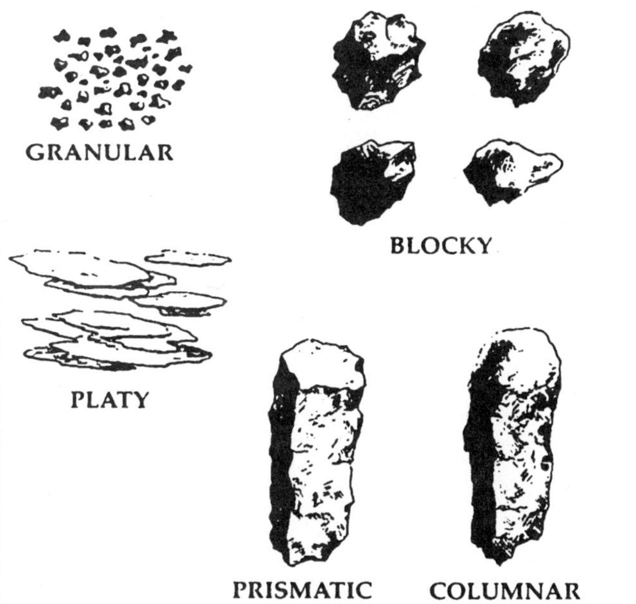

Choices of Soil Structure

- *Soil consistence* refers to how much the soil resists being broken apart. Pick up a single piece of soil and place it between your thumb and forefinger. Press until the piece of soil breaks apart. Use the following terms to classify the consistence of your soil:
  Loose: You have trouble picking up a single piece; it falls apart almost before you handle it.
  Friable: The chunk breaks with a small amount of pressure.
  Firm: The chunk breaks when you apply firm pressure but dents your fingers before it breaks.
  Extremely firm: You cannot crush the soil chunk with your fingers.

[1] Reprinted by permission of The G.L.O.B.E. Program, www.globe.gov. 1-800-858-9947.

- *Soil texture* refers to how coarse or fine the soil is. Take a sample of soil about the size of a golf ball and add enough water to moisten it. Mix with your fingers until it is uniformly moist. Then try to form a ribbon of soil by squeezing it between your thumb and forefinger. Use the following terms to classify your soil:

  Clay: The soil feels extremely stiff and sticky and requires a lot of pressure to form a ribbon.
  Clay loam: The soil feels sticky and is a little softer to squeeze.
  Loam: The soil is soft, smooth, easy to squeeze, and only slightly sticky.

  You will perform a more quantitative textural analysis in the next activity.

4. Before leaving the site, refill the hole by loosening adjacent soil and evening out the grade.

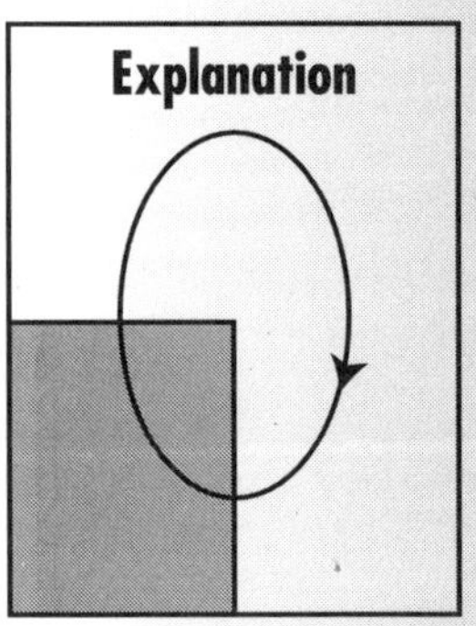

ACTIVITY 7.3

Laboratory Investigations of Soil

Student teams should perform these procedures with soil samples obtained in the previous activity and complete the data table for Activity 7.3.

Objectives

- Use laboratory methods to further characterize soil samples.
- Compare and contrast soil samples taken from different microclimates.

Soil Moisture Materials

beaker, balance, drying oven

Procedures

1. Measure the mass of a large beaker, and then add about 100 g of soil and measure again.
2. Dry the soil in a drying oven at 105°C for at least 10 hours, or allow to air-dry for at least 3 days.
3. Measure the mass of the beaker and dry soil, and subtract the result from the mass of the beaker and original soil sample to determine the mass of water in the sample.
4. Divide the mass of water in the sample by the mass of the original soil to determine the fraction of the original sample that was water.

5. Graph your findings to compare the results at different depths, or, using data from other groups, at similar depths at different types of sites.

Soil Texture Materials

100 mL graduated cylinder, 10% soil-dispersing reagent solution (10% Calgon® bath bubbles in water), distilled water, timer with second hand, flashlight; copies of the Textural Triangle on Student Log page 241

Procedures

1. Place 30 mL of dry soil in a 100 mL graduated cylinder. Gently tap the cylinder to settle the soil, and add more soil if necessary for a total of 30 mL.
2. Add 2 mL of 10% soil-dispersing reagent solution.
3. Fill the cylinder to the 90 mL mark with distilled water.
4. Cap or cover the cylinder tightly, and shake for 2 minutes. Be sure to turn the cylinder so all of the soil is mixed in.
5. Allow the cylinder to rest undisturbed for exactly 40 seconds. After 40 seconds, read the volume of sand that has settled in the cylinder, using a flashlight if necessary to see the settling line. Record.
6. Allow the cylinder to rest undisturbed for a total of 30 minutes. Record the volume of soil that has settled now, again using the flashlight if necessary. Record. Subtract the volume of sand from this total to obtain the volume of silt in the sample.
7. Subtract the total volume settling after 30 minutes from the volume of the original sample to determine the volume of clay in the sample.
8. Compute the percentage of sand, silt, and clay in the sample. Repeat twice, for a total of three times for each soil sample you wish to test.
9. Classify the soil texture using the Textural Triangle.

Textural Triangle

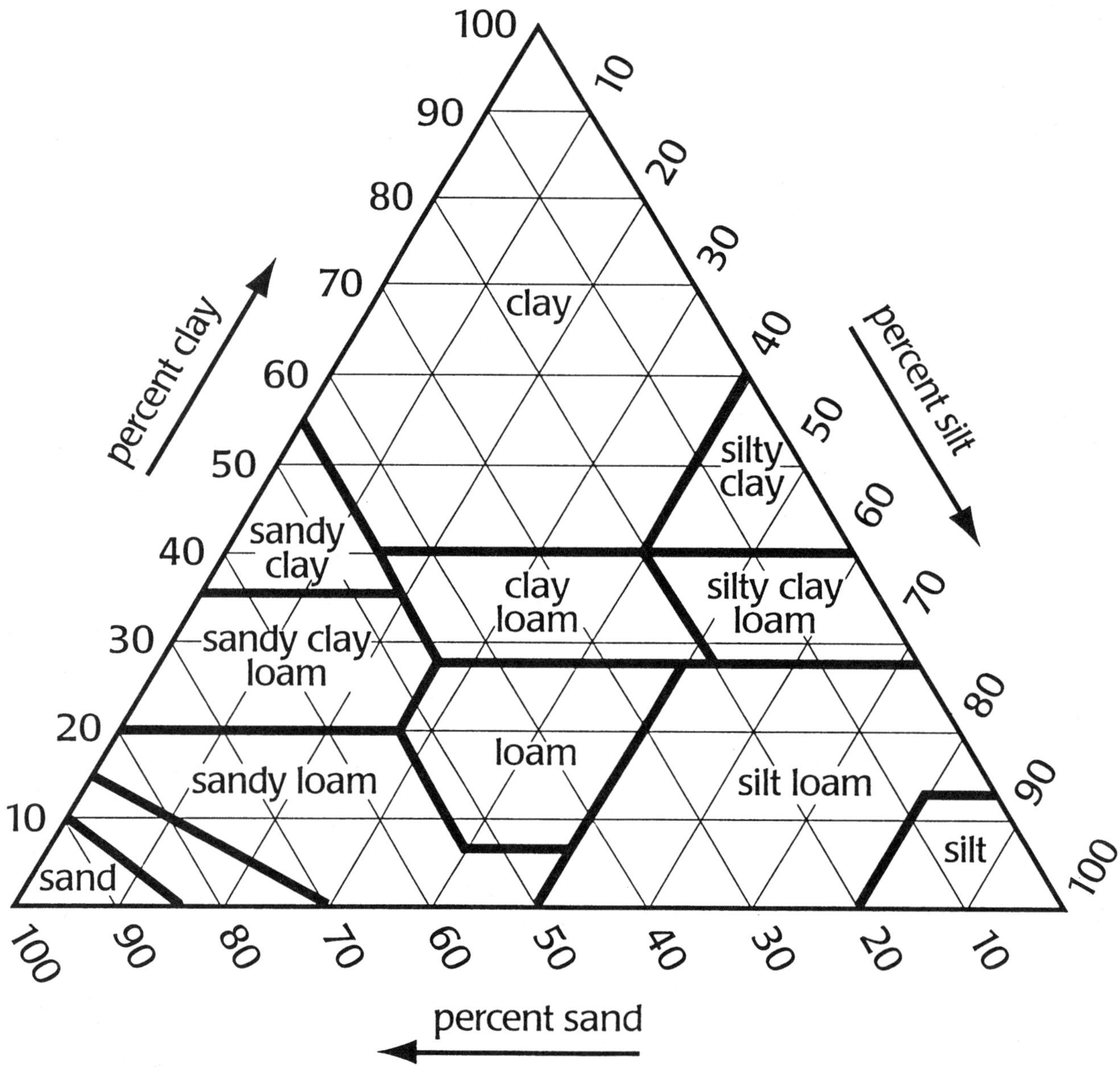

Students will use the Textural Triangle[2] to classify soil texture.

Soil Acidity Materials

small beaker, distilled water, graduated cylinder, pH paper or pH meter

---

[2] Reprinted by permission of The G.L.O.B.E. Program, www.globe.gov. 1-800-858-9947.

Procedures

1. Add 20 mL of distilled water to 20 mL of dry soil.
2. Mix thoroughly.
3. Determine the pH using a pH meter or pH paper. Record.
4. Repeat twice for a total of three times for each sample you wish to test.

Data Table Sample

Soil Moisture

Sample #	No. 1	No. 2	No. 3	No. 4
Mass of beaker				
Mass of beaker + wet soil				
Mass of wet soil				
Mass of beaker + dry soil				
Mass of water in soil sample				
% water in the original sample				

Soil Texture

Trial #	No. 1	No. 2	No. 3
Starting volume of soil			
Volume of sand settling after 40 seconds			
Volume of sand + silt settling after 30 minutes			
Volume of silt			
Volume of clay in original sample			
% sand			
% loam			
% clay			
Type of soil (from textural triangle)			

Soil Acidity

Sample #	No. 1	No. 2	No. 3	No. 4
pH (trial 1)				
pH (trial 2)				
pH (trial 3)				

Using the World Wide Web

Data Transmission

After the data have been compiled, teams should submit their data to the GTP Project Soil database. From the Project Soil home page, select "Data Sharing." Students will be able to enter their data directly into a Web form and send it electronically to the GTP database.

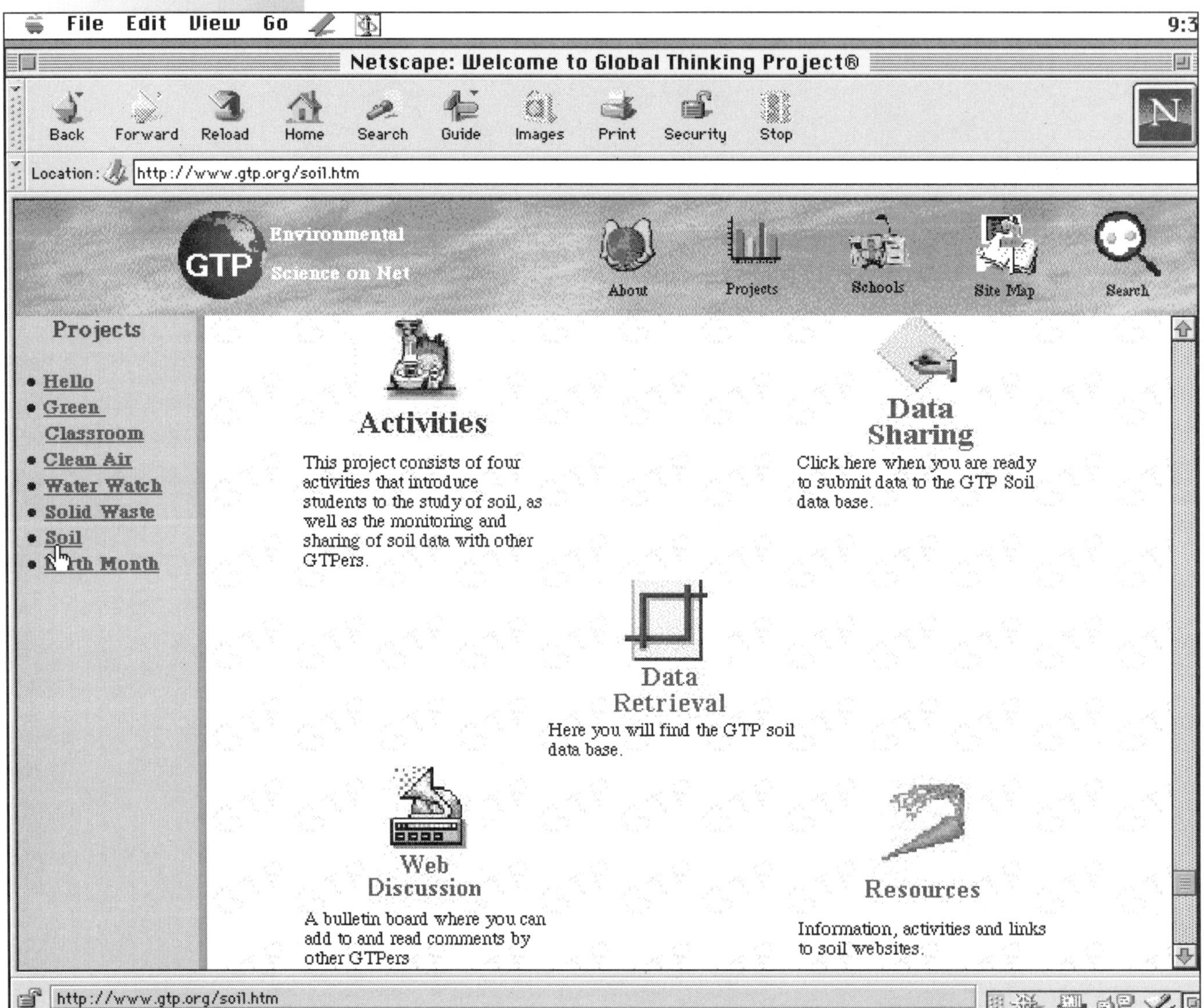

Project Soil Home Page

Data Retrieval and Analysis

From the Project Soil home page select "Data Retrieval." Allow each team time to obtain comparable data from a number of different schools. After a few minutes to study the data, ask teams to identify one or two

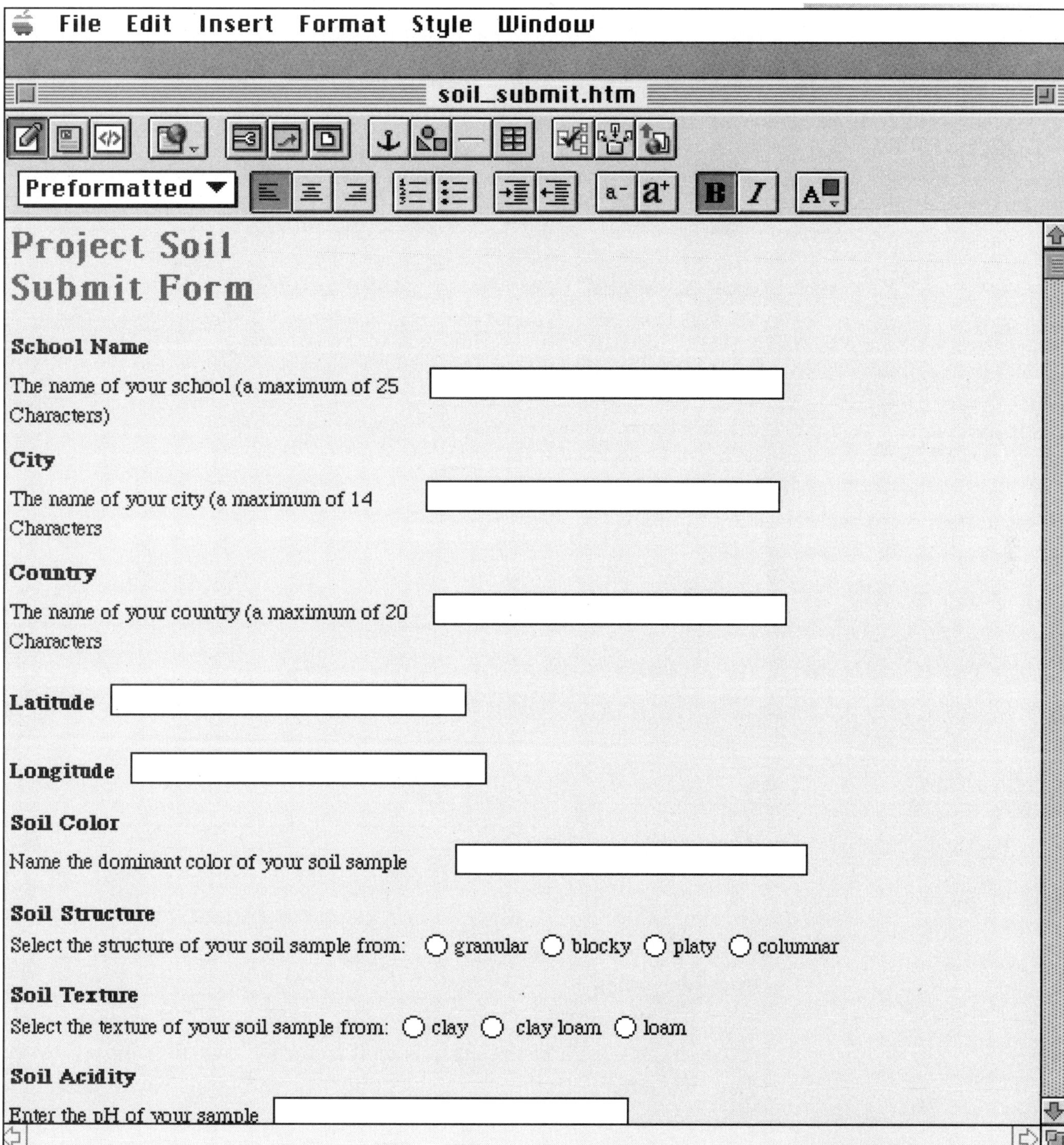

Project Soil Submit Data Form

soil properties for comparison across sites. Assist teams in formulating this comparison as a research question and in graphing or otherwise representing the data so that they can present their conclusions to the class.

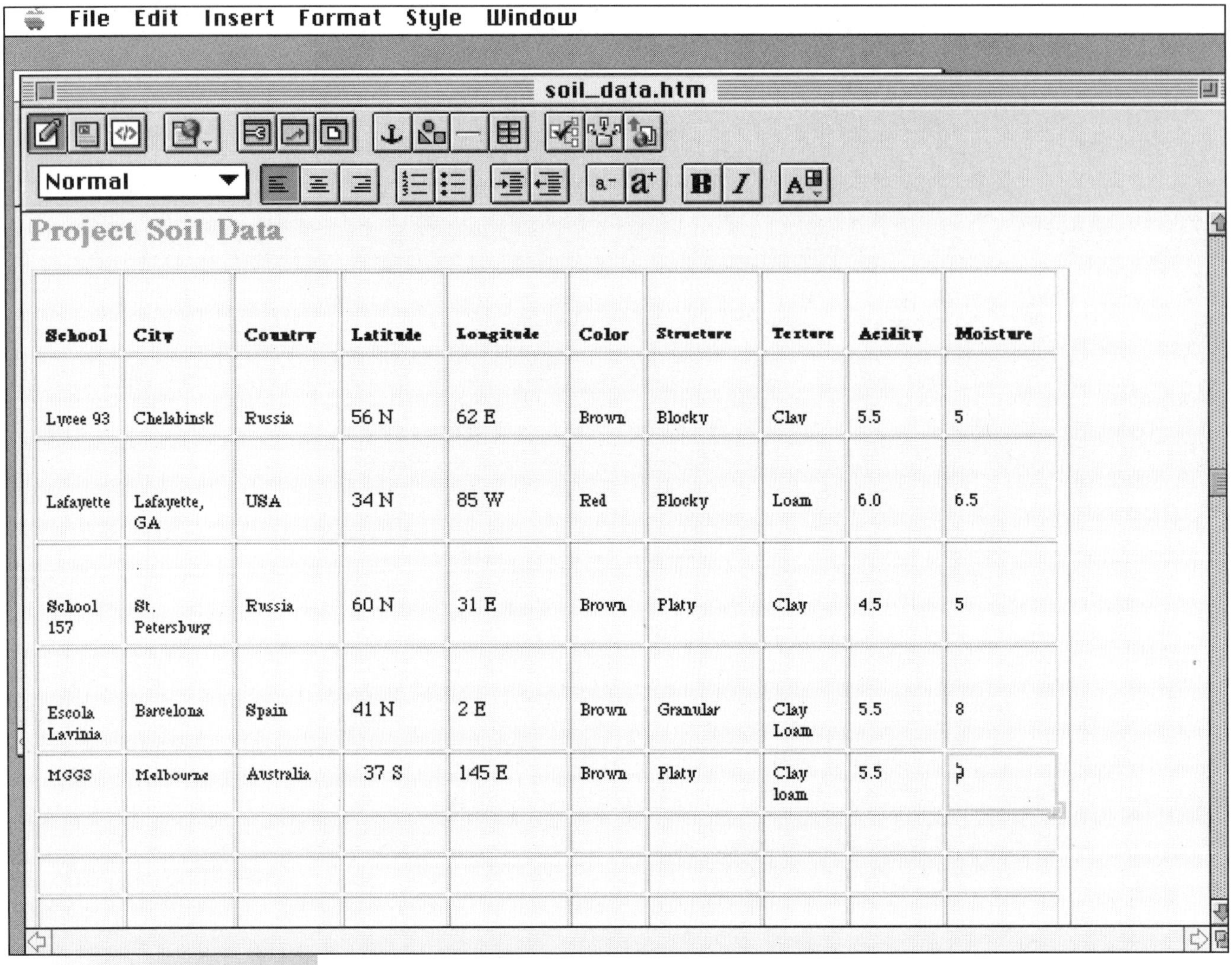

School	City	Country	Latitude	Longitude	Color	Structure	Texture	Acidity	Moisture
Lycee 93	Chelabinsk	Russia	56 N	62 E	Brown	Blocky	Clay	5.5	5
Lafayette	Lafayette, GA	USA	34 N	85 W	Red	Blocky	Loam	6.0	6.5
School 157	St. Petersburg	Russia	60 N	31 E	Brown	Platy	Clay	4.5	5
Escola Lavinia	Barcelona	Spain	41 N	2 E	Brown	Granular	Clay Loam	5.5	8
MGGS	Melbourne	Australia	37 S	145 E	Brown	Platy	Clay loam	5.5	5

Project Soil Home Page Data Table

Web Discussion

From the Project Soil home page, select "Discussion." Post each team's research question and findings, as well as further questions raised by the data. Respond to other teams' comments with your own ideas and questions.

ACTIVITY 7.4

Preserving the Soil

In this activity, students examine measures being taken in their community to preserve and enhance soil quality, and they implement a soil-enhancement project on the school grounds.

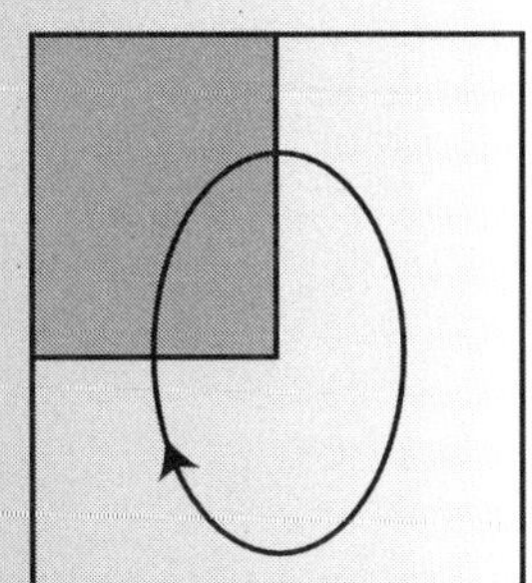

Taking Action

Objectives

- Discover measures being taken in the community to enhance soil quality
- Identify an area on the school grounds where the soil quality needs improvement
- Plan and carry out a soil-enhancement project

Materials

Materials for the project will vary; Student Log pages 242, 243

Procedures

1. Take a hike around the school grounds with your students. Are there areas of bare earth where there is evidence of erosion? Are there trees, shrubs, or flowers that look unhealthy? Discuss the importance of healthy soil in supporting plant and animal life. Why do the students think it is important to preserve and even enhance the soil?
2. Arrange to visit nearby sites where efforts are being made to preserve or enhance the soil. Visit a construction site, and ask about measures contractors take to preserve the topsoil. Visit a landscaper, and find out about xeriscaping. Visit a sanitary landfill and learn what measures are taken to prevent hazardous waste materials from leaching into the soil or ground water.
3. Back in the classroom, have teams brainstorm school-improvement projects they might take on. These might include such things as testing soil around unhealthy plants to determine whether specific plant nutrients are needed, designing plantings to help prevent bare soil from eroding, starting a composting project (and then using the compost to enhance the soil quality where needed), planting trees, or designing and building a wildlife habitat. Select one project as the focus for the class's activities.
4. Assist students in developing an action plan for carrying out their project. For detailed information about developing and implementing action plans with students, refer to Chapter 8.

CHAPTER 8

Project Earth Month

We, as fighters for the environment, will devote ourselves to protecting and preserving a healthy living environment. We hope to increase the world's knowledge of circumstances that plague our planet. If we all educate, demonstrate, and participate, we can make the people of the world "global thinkers."

GTP Declaration, Dunwoody High School and Moscow Gymnasium 710 Exchange Students, May 1997

During Project Earth Month, students will have the opportunity to design and implement environmental action projects.

Project Earth Month[1] is an open-ended project that encourages students to identify an important environmental topic, design investigations, and take action. In some cases students will write "declarations," as the students from Dunwoody High School and Moscow Gymnasium 710 did as a way to summarize their convictions after doing GTP research and working together on local environmental problems. These students believed strongly in the concept "Fighters for the Environment" as a motto of their devotion to taking action and being socially responsible youth. Under the guidance of their teachers, Trinna McKay at Dunwoody and Galina Manke and others at Moscow 710, these students demonstrated one outcome of an Earth Month project.

[1] A major contribution to this chapter was made by Susan Dunkerly-Kolb. She rewrote an earlier version of Earth Month, and we have modified her revision.

During the course of studying GTP, students have been learning how to investigate environmental problems and how to use the Internet to collaborate with students around the world. Students have also used the Internet to find information on environmental topics, as well as visit sites on the Internet that are created by scientists and teachers studying environmental change.

During Earth Month, which traditionally is held during the month of April in conjunction with Earth Day, the GTP invites your class to become involved in an action-taking project.

There are various types of environmental action strategies that students can take for Earth Month projects. Here are three possible approaches and examples of how American and Russian students, participating in a GTP exchange program[2] developed action statements describing how they would deal with one or more aspects of the environment. These action statements, called *declarations,* were presented by teams of ten American and ten Russian students at an Environmental Summit held in Atlanta by the GTP.

Persuasion: Convincing others, individuals and groups, that a suggested course of action is correct. GTP students, if they choose this course, must provide accurate data and ideas directed toward an environmental solution.

> *GTP Declaration:* At present pollution is increasing and the ozone layer is decreasing. We are not beyond hope, but if something isn't done, eventually we will face a crisis. We all play a part in this game we call "life." If we forget the rules, all the hope that still exists will disappear. We must work together to solve this global problem.
>
> We, students from Lafayette Middle School and Pushchino School 1, believe: Educating people about environmental problems is the key to solving problems. If you can change the way a person thinks, you can change the way he behaves!

[2] The GTP has supported exchanges of teachers and students among some of the participating countries in the GTP. Most recently the GTP, with grants from the United States Information Agency, has coordinated the exchange of more than 300 youth and 90 teachers between the United States and Russia. More information on GTP exchange activity can be found by visiting the GTP home page.

Political Action: Putting pressure on political or governmental agencies and/or individuals in an effort to influence positive environmental action. Students might draw up a charter that reflects the results of their research. You will read some charters written by teams of American and Russian students as they worked together on the GTP.

> *GTP Declaration:* The students of Yaroslavl and Salem High School believe that unless individuals take action, the environment will continue to deteriorate. Governing bodies should pass laws that make cleaning the environment more profitable than polluting it. The ideas of the Global Thinking Project should be promoted among the general population. Every person must make it his or her responsibility to protect the Earth.

Ecomanagement: Taking action to maintain a good physical environment or to improve a weakened physical environment. GTP students have learned through their research how humans can contribute to ecologically managing the Earth's important resources.

GTP Declaration: We, the below-signed students and teachers of Chattanooga Valley Middle School and Chelyabinsk Foreign Language Lycee 93 [who carried] out water and air monitoring in October 1996 until May 1997, state that:

1. Quality of water and air in Chelyabinsk and Chattanooga [is] different: The tropospheric ozone content in Chelyabinsk air sometimes reaches higher levels than in Chattanooga where it is practically nonexistent.
2. It rarely snows in Chattanooga and we could not compare the quality of melted snow in Chelyabinsk with that of Chattanooga. So we compared creek water. In Chelyabinsk water hardness is higher because of the presence of ions of calcium (Ca2+) and magnesium (Mg2+) in [the] water.

As the result of our participation in GTP we realized that there are many ecological problem[s] waiting to be solved. In connection with this our principles are

1. We are and should be citizen scientists and be optimists about the future of the world.
2. We should actively participate in solving the world's environmental problems.

In connection with the above, we declare that:

1. We want to be active citizens and help to solve environmental problems.
2. We want to help people realize that there are many ecological problems that need solving. For that we need to
   a) continue our water and air research
   b) do our ecological work with other groups (monitor the amount of waste and find ways to cut it down)
   c) use mass media to inform and persuade

The Earth Month project should be an opportunity for your students to apply the skills of doing projects, using the Internet, analyzing data, and collaborating in international research. The letter on page 151 is an example of an Internet report of an Earth Month project carried out by a group of fifth grade students in Rome, Georgia. Their project involved a number of activities, including a trip to the river they were studying during the year, as well as a presentation of their work at a local mall.

Dear Global Thinkers,

For Earth Month this year, we have been really involved with Water Watch (or Adopt-a-Stream, as it is also called). The creek we are studying is Silver Creek. We are trying to find out how the Pepperell Mill affects it, but also the continuing water quality of that site at the creek.

On Wednesday, April 17th, we went on our second field trip for Water Watch. First Randy Edwards, the head engineer for the mill, told us what the mill does, and how it has used the creek, over time. Second, we looked at a book that had old documents from the mill that dated back to 1895. We found out that the mill stopped using coal about 10 years ago when they learned about acid rain. Then we took a short hike along Silver Creek to see how other parts of the creek looked.

After that we did our normal testing at our adopted site. We tested the pH and dissolved-oxygen levels. We also caught macroinvertebrates, and today we are going to identify them.

We drove to the waste water treatment plant, and took a tour to see how the water from the mill was processed. We found that it was a lot cleaner when the water treatment plant got done with it. (The mill sends about 1/3 of the total amount of water that is treated daily to the treatment plant.)

Another thing we did for Earth Month was a setup in Mount Berry Square Mall. We took different ecosystems posters (like for a lake, mountain stream, wetlands, etc.) and showed how we use them in class. We had photos of kids in the field doing chemical and biological (macroinvert) testing. We showed our Global Thinking e-mail messages about Water Watch from earlier in the year.

We had a mural of our creek that we displayed, too. We brought macroinverts that people could come up and try to identify (with our help). We used the GTP ID sheets with this. We tried to show how art and poetry can be used to learn science. One of Mrs. Garner's favorite books, *The Water Planet,* was used. *In My Mother's House,* a book of poetry by Indian children, was also used.

On Thursday, we will finish identifying our macroinverts . . . we have one really weird one that we are not yet sure what it is. If we can't figure it out, we are taking it to Berry College. Dr. John Graham will know what it is.

Pepperell Elementary 5th grade,
Amanda Cole

Goals for Project Earth Month

1. Students will participate locally in action-taking projects on environmental issues or problems.
2. Students will identify an environmental problem or issue and design an action-taking project to help resolve the problem.

Planning Chart

Activity	Class Periods
Identifying an Earth Month project	1
Planning an Earth Month project	1–2
Taking action: Carrying out the design	2–3
Presenting Projects	1
Self-Evaluation: Ensuring long-term benefits	
Submitting a Web page to be linked to the GTP Web site	

Materials

Materials will vary for different projects. The Internet will be a central resource for the students doing Project Earth Month. The GTP home page for Project Earth Month contains a comprehensive set of links and project suggestions for students. It would be worthwhile starting Project Earth Month with a scavenger hunt to help students become familiar with the resources available to them. Students may need access to video cameras, multimedia experts, poster paper, drawing instruments, magazines, newspaper, phone books, and names of government agencies.

As an aid to locating reference books and organizations, a listing of references by topic area is located in Resources on pages 247–253.

Web Site Information

From GTP home page (www.gtp.org) click on "projects" to access Project Earth Month home page.

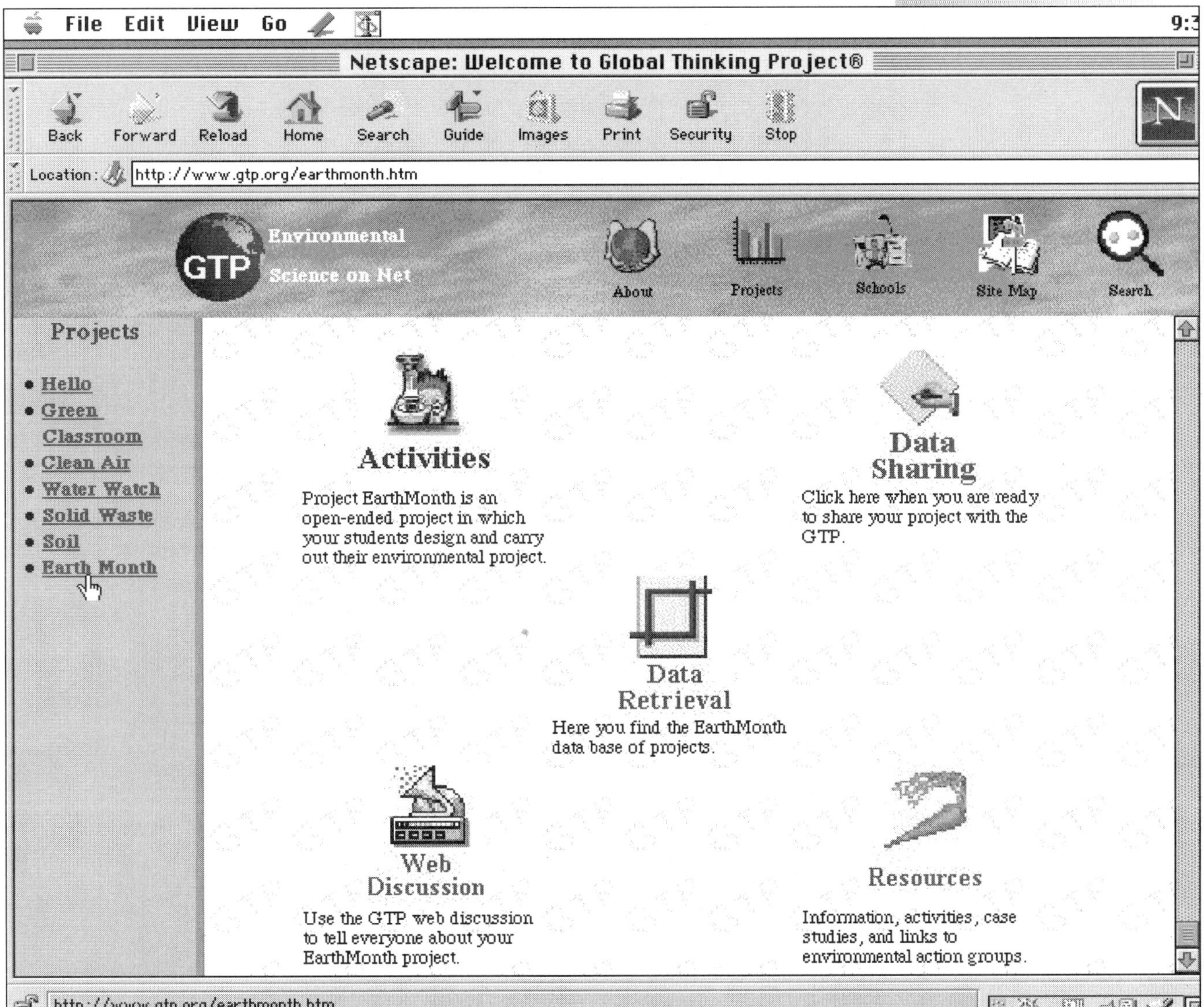

The Project Earth Month Home Page at the GTP Web site contains a comprehensive set of links and project suggestions for students.

Procedures

1. During the last week of March, conduct a brainstorming session with your students to identify possible Earth Month action-taking projects. Divide your class into groups, and provide each group with chart paper and writing tools to record their ideas for projects. Provide books and resources that students can scan to get ideas for environmental projects. Pool the groups' ideas onto a class Earth Month master table. Questions students should think about before deciding on an Earth Month project can be found on page 246 of the Student Log.
2. Follow the steps in the process outlined below to guide the students' work on Project Earth Month. The following steps will help them carry out their project:
   - Class determines subtopics and organizes into cooperative groups
   - Groups plan their projects
   - Groups carry out their projects
   - Groups plan presentation of their projects
   - Groups make their presentations
   - Students self-evaluate to ensure long-term benefits
   - Students submit a Project Earth Month Web page of their work
3. During the month of April, establish a schedule to help guide the students through their action-taking project. The table on page 155 outlines steps that you might find helpful.

Steps for Action-Taking Projects[3]

Stage	Description	Teacher's Role	Student's Role
1. Determine subtopics and organize into groups	Students scan sources, propose questions for an environmental project.	Introduce general topic of investigation; lead groups in discussion of topic and subtopics.	Generate questions of interest; sort questions into categories; form project groups.
2. Plan group projects	Group members plan project together; decide what they will study, and how they will go about it. They divide the work up among themselves.	Help groups formulate plans; meet with groups; help find resources and provide equipment and materials.	Plan what to study; choose resources; assign roles; send e-mail report describing plans.
3. Carry out projects	Group members organize and analyze information and data, pool their findings, and form conclusions. Members discuss their work in progress in order to integrate ideas.	Help with study and research skills; check with groups to maintain progress.	Seek answers to question; locate information; collect data; interview people; integrate findings and summarize.
4. Plan presentations	Members determine the main idea of their presentation and how to present their findings.	Discuss with class plans for presentations; organize the process.	Determine the main idea for the presentation; plan how to make presentation; prepare materials for presentation (poster, video, etc.).
5. Make presentations	Presentations are made to class and others.	Coordinate presentations; insure that e-mail reports are sent to Global Community and gtp.earthconf.	As presenters, give feedback to classmates; send e-mail reports; read and react to e-mail reports from other schools.
6. Evaluate	Students share feedback about the projects. Projects are evaluated by teachers and students.	Evaluate learning and summarize the projects.	Reflect on work as a group member; reflect on impact of project on environmental literacy.

3 Based on *Expanding Cooperative Learning Through Group Investigation* by Yael Sharan and Shlomo Sharan. New York: Teachers College Press, 1992.

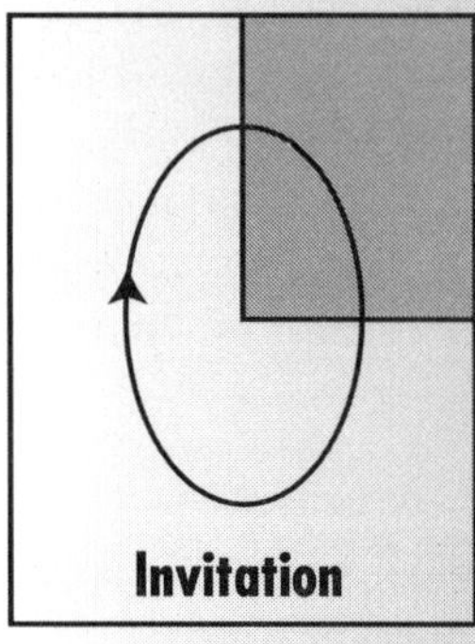

ACTIVITY 8.1

Determining Subtopics and Organizing into Groups

Objectives

- Learn how to explore environmental resources.
- Be able to brainstorm possible questions.

Procedures

1. Begin the brainstorming process by reflecting on the project up to this point. Ask the students to think about what they have learned from their own environmental monitoring activities and from communicating with other schools about their environmental problems. Have a discussion with students about which environmental problems they consider most important, and why. Share with your students your own personal concerns.
2. Conduct an Internet search session with the students. You might want to use the scavenger hunt that is located on the Project Earth Month Web page as a guide for their exploration.

   Provide students with resources that allow them to answer some of the questions they may have about their environmental concerns. You will find resources in your media center or your local library; perhaps, also, you have collected your own personal set of resources. We have provided a list of resources at the end of this book for your use.
3. Once your students have had the opportunity to inform themselves, begin brainstorming a list of ways they might want to act on what they have learned. You might want to discuss with them the various kinds of environmental action listed in Activity 8.2. Once they get an idea of their individual areas of interest, help them divide into project groups based on those interests.

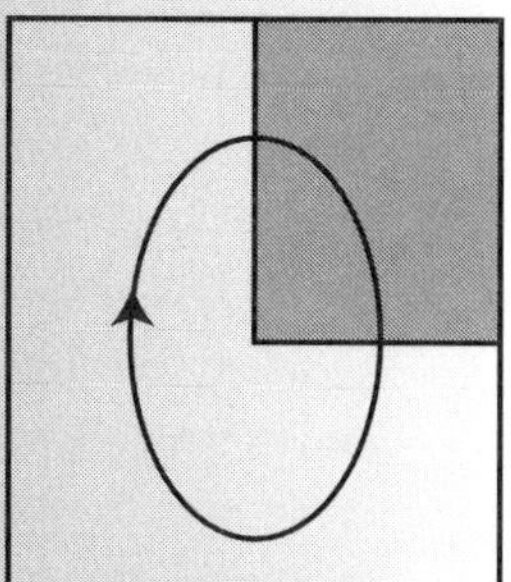

Invitation

ACTIVITY 8.2

Groups Plan Their Projects

Objectives

- Define the nature of students' Earth Month Projects.
- Formulate an action plan to carry out their projects.

Procedures

1. Groups plan projects: Once the students have gotten into their project groups, they can begin to define their project. There are many ways to design an environmental action project, and by keeping in mind the different learning styles, interests, and personalities of your students, you can guide them towards an appropriate choice. For instance, some students may have interest in the area of art, music, or writing, while others may want to do a more scientific investigation. The projects should help students make a personal contribution to the preservation of the environment. This will make the experience a longer lasting one for everyone.

   Be sure to involve parents whenever possible in the design, execution, and presentation of the projects. This provides a family aspect and also aids in educating the community about student environmental concerns.

   Following is a list of possible projects. On looking at the list, you may be able to come up with other areas and other kinds of projects. Use this list to guide your students' choices, rather than providing it as a summary of the only choices available. (These suggestions came from a brainstorming session done with GTP teachers during the 1994 Summer Institute and from an environmental sourcebook called *Environmental Education Teacher Resource Handbook: A Practical Guide for K–12 Environmental Education,* edited by Richard J. Wilke (Millwood, NY: Kraus International Publications, 1993). Ordering information can be found in Resources.

Research Projects

- Measure and report on solid waste collected at your school for one day.
- Have an "Invention Convention" on new ways to reuse things.
- Restore a waterway bank.
- Monitor and report on water quality of a local river, stream, or lake.
- Monitor and report on air quality in a particular area.
- Conduct acidity tests on rainfall in your area.

- Perform a wildlife population study.
- Inventory trees or other plants in the schoolyard or nearby natural area.
- Conduct surveys to assess students' or parents' behavior and attitudes about an environmental topic or issue.
- Survey patrons of a local grocery store to determine how many people are buying environmentally friendly products.
- Determine how many people are carpooling in the community, and why or why not.
- Conduct energy- or water-use audits within the school or certain households.
- Investigate methods of organic pest control.
- Test the biodegradability of various products.
- Conduct an experiment to determine the resistance of various plants to acid rain.
- Simulate population growth studies using yeast, insects, or duckweed.
- Design, construct, and test solar-powered ovens or other devices.

Construction and Beautification Projects

- Adopt a stream or highway.
- Plant trees or flowers.
- Build bird or bat houses.
- Develop school, home, or community composting facilities.
- Build an environmental learning center, such as an outdoor classroom.
- Turn a school courtyard into a nature study center.
- Construct a display case that contains materials on natural history, local and national environmental issues in the news, and highlights of students' environmental projects and activities.
- Clean up litter in community areas or the schoolyard.
- Plant trees.
- Improve habitat for wildlife through schoolyard landscaping that uses native plants.
- Design or clean up a nature trail.
- Plant a butterfly garden.
- Plant a school or community organic garden.
- Clean up a local pond.

Projects That Utilize Art or Language Arts

- Put together a collection of "Environmental Tip of the Week" suggestions for each week of the school year, to be read during morning announcements.
- Have a group put together a weekly "Eco-Report" that gives the environmental news of the week, to be read during the morning announcement or included in the school newspaper.
- Put together a Global Thinking Project display, describing the project and the schools involved.
- Have a poetry or rap contest.
- Have a photo essay on an environmental theme.
- Create an art display of art made entirely from recycled materials.
- Write and present speeches during a school event, such as Earth Day.
- Create and perform a play or skit to encourage people to act responsibly toward the environment. Perform it for the school or community.
- Design posters and flyers to encourage people to support certain environmental positions or take certain actions, and have a poster contest.
- Conduct a debate in school about an environmental issue.
- Write a letter about an issue to the editor of a newspaper.
- Submit written viewpoints that could be included in a newspaper article or read on a television or radio news program.

Projects to Raise Money for the Environment

- Design and sell T-shirts with environmental themes.
- Have a recycled-art sale.
- Conduct a film festival on environmental issues.
- Conduct a fund-raiser and sell environmentally friendly products, such as recycled paper, wildlife T-shirts, organic snacks, local crafts.

Political Action Projects

- Phone or write letters about an environmental issue to an elected official.
- Conduct a "get out and vote" campaign to support an environmentally responsible candidate.
- Speak about an environmental issue at a public hearing.

2. Helping Students Develop Their Projects: Your students will have all sorts of ideas for projects, and it will be important for you to help them determine which ones are appropriate and feasible. Below is a list of questions to ask yourself when reviewing student suggestions.
   - Did the students come up with this idea on their own and can they carry it out on their own?
   - Does the project make sense in terms of both student concerns and the overall objectives for the class?
   - Are the students realistically considering the time factor involved in conducting this project?
   - Will the project encompass the various learning styles and interests present in the project group? If not, could some aspects of the project be changed, or could something else be added?
   - Are there interesting interdisciplinary aspects involved or could there be?
   - Can you evaluate the project appropriately?
   - Can you foresee any potential liability problems associated with the project?
   - If the students are going to have to leave the school grounds, what kinds of parental permission will be required?
   - Are any special arrangements going to have to be made such as for use of equipment and special materials, transportation, use of certain facilities, guest lecturers, or parent volunteers?
3. Helping Students Work Cooperatively: Once your students have formulated their projects, they will need to assign tasks. Keeping in mind the abilities and personalities of each group, you can help them with this process. The cooperative learning principles of positive interdependence, individual responsibility, face-to-face interaction, and small-group collaborative skills will come into play here, as they have all year as students have worked through the various projects of the GTP.

   Help each group develop an identity. Perhaps the students may want to come up with a name for their group and to develop a group emblem or banner. You can help students be interdependent by ensuring that each member has an important part of the entire group's task to complete. To ensure individual responsibility, you might have the groups build checkpoints into their project plan so they (and you) can periodically check in with one another to see that each person is where he or she is supposed to be. Build into the class period an opportunity for the group to work together on the project, rather than having each student work individually. Interaction with one another will provide the students with opportunities for support and modeling of learning behaviors that they will not be able to obtain alone. You may want to go back over the some of behaviors that allow cooperative groups to work more smoothly, such as: checking to make sure everyone in the group understands directions; seeing that everyone learns basic material before going on to new information; encouraging everyone in the group to participate; listening attentively to what other group members are saying; and criticizing ideas, not people.

4. Formulating Student Plan of Action: Student Project Planner. By helping your students to develop a realistic plan of action, you will ensure that they have a better working relationship with one another, an easier process, and a better product. Have students put together a "suspense" calendar on which they lay out the tasks they will need to accomplish in order to come up with a satisfactory project. In creating this calendar they will want to include deadlines for accomplishing tasks, such as the ones below, plus personal and class deadlines that may affect the outcome of the project.

Project Planning Tasks

- Formalize our project goals and objectives: What do we hope to accomplish with this project? How will we know if it has been a success?
- Determine our research needs, and divide these specific tasks among group members.
- Determine what kinds of materials, special equipment and other resources will be needed.
- What are the monetary costs, if any, involved in the project? How will we deal with these costs, and how can we minimize them?
- What special arrangements need to be made ahead of time so they will be in place when they are needed? (Consider transportation, use of computers or other special equipment, use of a particular room in the school, scheduling a guest speaker or publicity event, and so on.)
- When and where can we meet to work on our project? If there are to be any meetings outside of school, can everyone be there?
- Determine a method of keeping track of our progress (journals, photographs, videotapes, etc.). What do we want to include in this account?
- Determine how, when, and where we are going to present the results of our project to others.
- Develop a workable outline for the project, including all the tasks necessary and who will be responsible for carrying them out.

Taking Action

ACTIVITY 8.3

Groups Carry Out Their Projects

Objectives

- Act on students' action plans.
- Collect data, make observations, keep journal/log notes, and create (or "develop") a portfolio.

Procedures

1. Now that the students have developed their plans of action, they will begin the actual work of carrying out their projects. During this time, students will be organizing and analyzing the information they have gathered during their research and using it to carry out whichever kind of project their group has chosen. Your job is to be a facilitator of this process. Your tasks fall into two main categories: providing assistance with the actual project itself, and helping the project groups work together cooperatively.

   When providing assistance to groups, the degree of your involvement will, of course, depend on the motivation and skill levels of the students within a group and the nature of the project they have chosen. Some students will possess more motivation and skill than others; however, all will have a better chance of developing the skills and motivation they need if they are allowed to develop the inquiry themselves, rather than being told what to do and how to do it. As with any project, some risk is involved, and students must, within limits, be allowed to try things you know may probably fail. It is in dealing with these occasional wrong decisions and deadend inquiries that students learn from their mistakes and develop the critical thinking skills we talk so much about. During your monitoring of the project groups, you will wish to assist in such areas as clarifying instructions, reviewing important procedures and strategies for completing the assignment, answering questions, and teaching task skills as necessary.

   As with the other projects we do in the GTP, cooperation is key. Your other task as facilitator of the project groups will involve possible intervention in order to stress collaborative skills. Each group should have two main goals: to accomplish the task successfully, and to work together well as a group. Often, project groups will be so task oriented that they will neglect the maintenance of their group relationship. Work sessions within the groups should be pleasant, lively experiences. If no one is having fun, something is going wrong. Problems with collaboration should be brought up and solved before they get out of hand, and continuous effort should be made to maintain the group's positive relationship.

Deciding when and when not to intervene in a group problem is part of the art of teaching, and by using a little restraint, teachers can usually learn to trust their intuition. Even when intervention is necessary, teachers can often turn the problem back to the group to solve. If you observe a problem, you may want to have group members momentarily set aside their tasks and ask them to come up with an adequate solution to their problem. Remember, the best time to teach cooperative skills is when students need them.

2. Teacher Evaluation of Process and Projects: If you plan to grade this project as part of a class assignment, you will want to give some thought beforehand to the criteria upon which you will base your evaluation. Goals, objectives, or learning outcomes for the project can be used as references for evaluation. Since so much is learned during the process of such a project, the quality of student participation, more than the final product, should be the main consideration for evaluation. Even if your students are participating in this project as an after-school type of endeavor, they should give some thought as to how to evaluate their progress. There are a number of ways students and teachers can evaluate student progress.

Observations

Certain factors can indicate the level of student involvement in the project. Attendance, enthusiasm, and participation are all aspects to consider during observation. Observe also the physical development of the project itself, its rate of construction, and the quality of its design.

Journals and Portfolios

Journals allow students the opportunity to reflect on the process of carrying out the project. Journal assignments can center on such ideas as that of having students keep track of their progress, evaluating the effect the process has had on them as science students and environmentalists and how the project is helping the environment. Teachers can have students put together, in a portfolio, artifacts documenting the project's progress. Artifacts can include photographs, rough drafts of reports, presentation materials, and evaluations. By having the students choose what to include in their portfolios, teachers can get an idea of what aspects of the project students consider themselves to have done well and which aspects they consider most important.

Reports and Presentations

Students will be writing and e-mailing a project plan and project results to their global community and the GTP listserve. If you set up clear objectives for these reports, they can be used as evaluative pieces. Objectives can include such criteria as development of the idea, clarity (especially when being composed for an international audience), and inclusion of important points. Presentations of the projects will be given to classmates and possibly the school, community, and other GTP classes. You may want to have the students do pilot presentations for another group, during which they would be evaluated by peers using a set of criteria developed by the class. If you use this method, the class should have the opportunity to develop the criteria and have access to them to use as a framework for developing their presentation plans.

Explanation

ACTIVITY 8.4

Groups Plan and Make Presentations

Objectives

- Learn to prepare material for a presentation.
- Participate in an Earth Day Celebration.

Procedures

One of the most important aspects of environmental action is education. Daily, we make decisions which affect the environment, and so our ability to make informed decisions is of extreme importance. Presentation of the students' Earth Month projects allows them to educate their classmates and possibly the school, community, and other students involved in the project. Below are some suggestions for allowing the world beyond your classroom to benefit from your students' efforts.

1. Earth Day Celebration: Traditionally, Earth Day falls on April 22. Special events are held during the entire week of April 22, which is designated as Earth Week. Although environmental education should occur during the entire year, a day-long or week-long event gives the students an opportunity to learn in a more in-depth way about one or a number of environmental topics. One of the real benefits of a celebration of this type is that it provides an opportunity to involve administrators, parents, university professors and students, and the community.

   Check with other teachers in your school. They may already be planning some event for Earth Day or may want to help you put together something fun. Involving other teachers will allow different classes to take responsibility for different parts of the event and therefore share the responsibility and inject different ideas. An Earth Day event will be the perfect opportunity for your students to showcase their Earth Month projects. Provide time during the event for your students to present their projects. They can present to the entire school or their grade level in a central area, or perhaps visit various classrooms during the day or over several days.

However you choose to conduct your celebration, be sure that your students are truly involved in the process from planning to cleanup. In this way they can more clearly understand the reasons behind their participation, and the event will have more than a one-day impact.

Note: Be sure to visit the Project Earth Month home page for the latest information on Earth Day.

2. Publicity for Your School and the Environment: Publicizing your students' Earth Month projects has a number of benefits. By making the school and community aware of the projects, you develop awareness of the environmental problems that have been under study. Publicity also encourages others to recognize the students' hard work, and this can be very gratifying. Particularly if any of your students' projects has been designed to benefit the community, the public should be notified that it is taking place. Administrators, parents, and other members of the community can learn about the projects by being invited to participate in them, from their inception through their presentation. If you have others outside the class involved in the projects, publicizing the presentations and inviting the participants to view the final products are ways of recognizing their efforts.

   Some ways you can publicize the presentations of your Earth Month projects include

   - Having students design and hang posters or distribute fliers
   - Including a write-up about the project in the school or parent bulletin
   - Having students write and send out press releases to local newspapers and radio and television stations
   - Asking a local radio station to do a public-service announcement
   - Inviting the media into your class to do a human-interest story

3. Student Awards: As part of your event, you may want to include an awards celebration. Recognition of students' efforts ensures that they will remember their project for years to come. You may want to design the awards yourself, or the administration might have some ideas for recognizing your students' efforts.

Taking Action

ACTIVITY 8.5

Student Self-Evaluation: Ensuring Long-Term Benefits

Objectives

- Reflect on the Earth Month activity and consider what was learned.
- Evaluate the impact of the project.
- Use the GTP Internet resources to collaborate and communicate with others concerning student projects.

Procedures

1. Once the excitement of project presentation is over, students need the opportunity to reflect on the project experience. Careful planning of particular aspects of the project will ensure that later reflection and evaluation creates a rich experience. Self-evaluation requires reflecting back to the project goals and objectives. Were these goals and objectives met to the students' satisfaction? If not, why not? What would the students change if they had the opportunity to repeat the experience? What did the students feel they learned about their environmental concerns through this experience? What did they learn about environmental action? How can students see themselves remaining involved with this problem?
2. A very important set of questions for students to explore involves their perception of how their project impacted their environmental concern. Students should begin to ask themselves these questions from the earliest planning stages. Global discussion via e-mail with other GTP schools will also provide information. Students might consider asking other GTP schools to comment on their project plans and presentations. In addition, students should expose themselves to information from other schools about their own projects. This will provide them with ideas they may not have considered. By exploring the work done by other schools, new questions concerning the impact of their own projects will develop.
3. The use of journals and portfolios will greatly enhance the self-evaluation process. Reflective questions posed to students throughout the project development process will guide them along the path toward final evaluation of the project. Peer review and evaluation of portfolios during the process will also allow for easier compilation of the final product.

ACTIVITY 8.6

Developing an Earth Month Web page

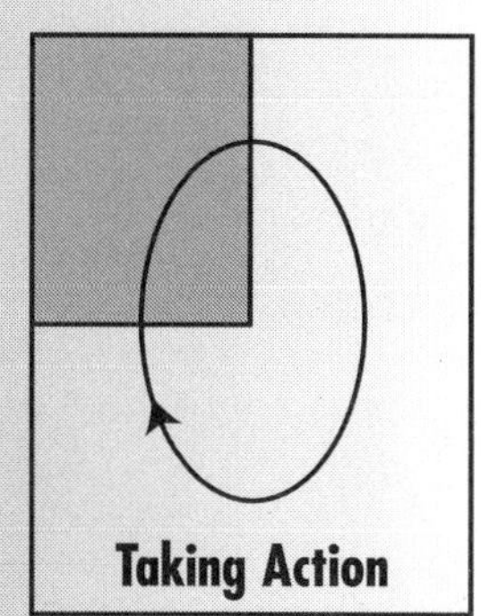

Objectives

1. Develop a Web page summarizing the results of research
2. Submit the Web page for publication on the GTP Web site

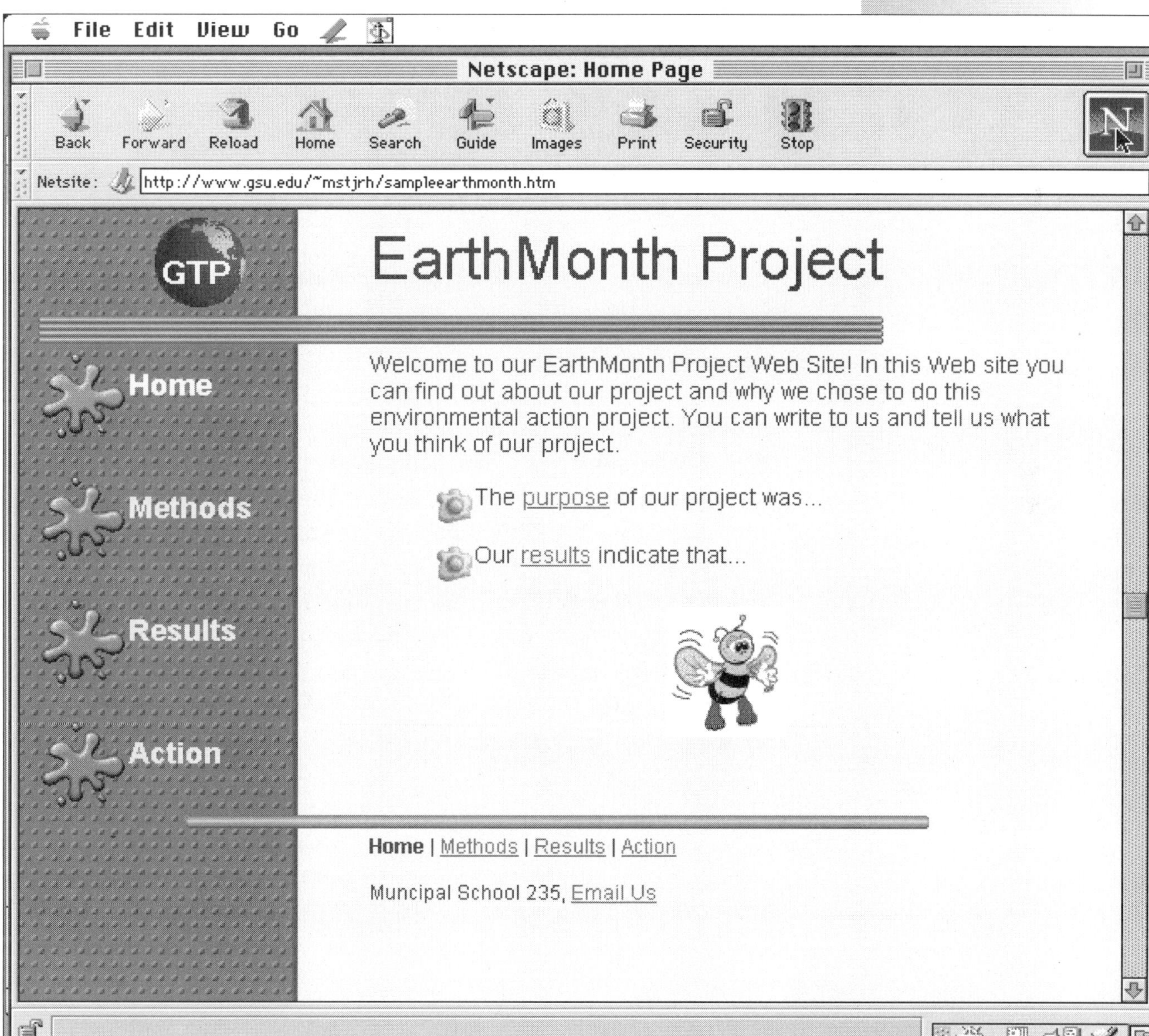

Web Page Template for an Earth Month Web Page

Student projects will be developed in many different forms, using a variety of strategies and media. It is the intention of the GTP, however, to document this work and to celebrate its results with teachers and students around the world. Thus, we are organizing a campaign to publish on-line the results of the students' work. Results will be published on the Global Thinking Project Web site.

The Web-page template shown on page 167 might give you and your students some ideas for development.

The Web pages that you develop can include:

- Text: Students should describe their project, including the problem they studied, the results and their conclusions, and the actions that they have taken.
- Data: Students should include data that was relevant to their project, how they collected the data, an analysis of the data, their interpretation of the data, and the conclusions they made.
- Photographs: Students can include photographs within their Web page.
- Drawings: We encourage original student drawings. You can scan the drawings and insert them into your Web page.

Data Sharing: When your students have their Web page ready to be linked to the Project Earth Month Web page, send a message to the GTP from the Earth Month Web page.

Web Discussion: As other schools develop and have their Earth Month page linked to the Earth Month home page, have your students browse the pages and then write brief statements to add to the Web discussion on the Earth Month home page.

Video Conference: You might be interested in bringing together other GTP schools to discuss the results of their Earth Month projects via a video conference. You can coordinate a video conference yourself, or ask the GTP to help you with logistics. Visit the GTP home page for information about video conferences.

CHAPTER 9

Extension Activities for Global Thinking

Extension Activities are Global Thinking lessons developed by teachers that you might want to consider as alternative activities for your students. In some cases, the lessons can be accomplished within your class to enrich your students' knowledge of global thinking. In other cases, your students might want to invite other classes to join with them to accomplish the activity. The activities are:

Activity 9.1 Exploring Attitudes Toward Other Nations and Peoples
Activity 9.2 Imagining My Future
Activity 9.3 Water, Water Everywhere

ACTIVITY 9.1

Exploring Attitudes Toward Other Nations and Peoples[1]

(Use with p. 23 in Chapter 2.)

Contributed by: Dorothea McAlvin
Mt. Zion High School
Jonesboro, Georgia, USA

The following activity provides an opportunity for student teams to explore, assess, and compare their perceptions of foreign nations and peoples, using one version of a survey instrument developed for a UNESCO-sponsored study in the late 1950s. (Lambert and Klineberg, 1967) As a culminating activity, students compare and contrast their results with those of the original study, as well as with a 1987 follow-up study.

[1] This activity is used with permission of Dr. Dorothea McAlvin.

Objectives

- Formulate generalizations about perceptions of foreign peoples.
- Formulate generalizations about influences on perceptions of foreign peoples.

Materials

1 copy of interview questions per pair or triad
1 copy of UNESCO survey per team

Procedures

1. Open the lesson by asking students to name foreign nations that are similar to yours. (Stop after 3 or 4 names.) Continue by asking how those nations are similar to yours.
2. Briefly give the background of the UNESCO-sponsored study. (Refrain from reporting the results of the study.)
3. Divide student teams into pairs or triads. Give each pair or triad a copy of the interview questions from page 173. Have each team member interview the other, and record the answers.
4. When students have finished interviewing each other, have them return to their teams to consider the following questions:
   1. Which countries were named as most similar to ours?
   2. Why were they characterized as similar to ours?
   3. From what sources have we acquired our knowledge of these countries?
   4. Which nationalities did we choose to be? Which nationalities did we not want to be?
   5. From what sources have we developed our attitudes about foreign peoples?
5. Provide teams with results from the 1959 and 1987 studies in the UNESCO survey. Questions to consider in teams and then as a whole class include the following:
   1. To what extent do the answers given in 1959 reflect that time period?
   2. What are some of the events that determined people's reactions to foreigners in the 1950s?
   3. Based on the 1987 study and your own data, what answers have changed since 1959? Why do you think they have changed?
   4. If the same questions are asked 10 years from now, what changes do you predict?
   5. What recent developments would tend to change people's attitudes toward one another, and in what way?
   6. What new sources of knowledge and attitudes are emerging today?

Results of the 1959 UNESCO Survey, and 1987 Follow-up

In the late 1950s, a UNESCO-sponsored survey was made of 3,000 children in 13 ethnic groups in 11 nations around the world to examine their views of foreigners. Data from the interviews, conducted in 45-minute sessions, were analyzed by Wallace E. Lambert and Otto Klineberg and published in their 1967 work, *Children's Views of Foreign Peoples. A Cross National Study* (New York: Appleton-Century-Crofts).

Children in the U.S. (white and black), Canada (English-speaking and French-speaking), Turkey, Lebanon, Japan, West Germany, England, France, South Africa (white and black), Lebanon, Italy, and Israel, ages 6, 10, and 14, were asked their opinions about foreigners, the source of their opinions, which foreign people were most like themselves, and which were most unlike themselves. In addition to any countries they might have named spontaneously, the children were asked specifically about seven reference peoples: Americans, Brazilians, Chinese, Germans, Indians (from India), black Africans, and Russians.

American children were found to have positive attitudes toward foreigners. They seemed to become increasingly positive between the ages of 6 and 10, and more negative between the ages of 10 and 14, but they still had positive attitudes at age 14. American children viewed their country as "wealthy and free." They found it difficult to name another nationality "like" themselves, and less difficult to name another nationality "unlike" themselves. They named British, Canadian, French, Italian, and German as nationalities similar to Americans, and African, Chinese, Japanese, and Russian as nationalities that are dissimilar. Older children increasingly named the British as a similar nationality and the Russians as a dissimilar nationality. American children were more negative toward Russians than toward any other group, though more than half of the American children even at age 14 were prepared to view Russians as "friendly" rather than "unfriendly."

As part of a research project in 1987, McAlvin adapted the questions from the Lambert and Klineberg study and used them to examine the opinions of students (ages 14 and 15) in a public high school in a predominantly middle-class section of suburban Atlanta. Fourteen students were randomly selected from the 96 students in their age group who had been part of the larger study. The following is a summary of the opinions expressed by this group of students:

1. Students did not feel responsible for knowing about any country they had not studied in school.

2. The British were viewed as most similar to people in the United States, and British was the nationality these students would like to be if they were not Americans.
3. Africans and Soviets were viewed as the people most different from people in the United States.
4. Students tended to base their evaluations on their perceptions of the degree of political freedom, civil rights protection, and material well-being enjoyed by the people of a given country.
5. Unlike children in the 1959 survey, these students found Japan to be "like" the United States.
6. Most of the students (13 of the 14) named "Soviet" as a nationality they would not like to have. They cited governmental repression, enmity between our two nations, Soviet military power, the difficulties encountered by consumers in a planned economy, and mobilization of opinion ("The way they talk about us, and the way we talk about them.") as reasons for their evaluation.

Interview questions adapted from a UNESCO-sponsored survey

Interview Questions

1. Name a country where there are people who are *like* you or are *similar* to you.
   - In what ways are they like you?
   - What else do you know about them?
   - Do you like them?
   - Why do you say that?
   - How do you know about them?
2. Are there any other people who are *like* you? (continue with the questions given above)
3. Name a country where there are people who are *different* from you.
   - In what ways are they different?
   - What else do you know about them?
   - Do you like them?
   - Why do you say that?
   - How do you know about them?
4. Are there any other people who are different from you? (continue with the questions given above)
5. Now ask the same questions about any nationality or nationalities in which you might be interested.
6. If you were not [your nationality], what nationality would you most like to be?
7. If you were not [your nationality], what nationality would you least like to be?

Optional Extensions

1. Have students share interview data with students in other types of communities in their own countries, and with students in other countries, or
2. Have students write their own interview questions to be used to interview their parents, or students at a different grade level, in another community in their own country, or in another country. Report results in the GTP listserve. Questions to consider might include the following:
   - What answers are similar to yours? What answers are different?
   - What generalizations can be drawn about (a) people's knowledge of other people, (b) people's attitudes toward other people, (c) differences in knowledge and attitudes that seem to correlate with age, (d) the different uses made of sources of information about the world?

ACTIVITY 9.2

Imagining My Future

(Use with p. 31 in Chapter 2.)

In an introductory activity to the Student Study Guide for *Biological Science, An Ecological Approach,*[2] students are invited to predict what their futures will be like. We include the questions from this activity here, to encourage your students to begin to examine the implications of the content they will study as part of the Global Thinking Project on their future lives. These questions also provide an opportunity for students to conduct cross-cultural research on their perceptions of the future.

Objective

- Predict what the future might be like.

Materials

Predicting Your Future activity questions, page 175 (1 copy per student)

Procedures

1. Hand out the Predicting Your Future activity questions and ask students to complete them without consulting one another.

[2] Adapted from *Biological Science Curriculum Study, Student Study Guide for Biological Science, an Ecological Approach*. Dubuque, IA: Kendall/Hunt Publishing Company, 1987. pp. 1, 2.

Predicting Your Future Activity Questions

Predicting Your Future

The purpose of this activity is to encourage you to imagine what you think your life will be like when you are the age your parents are now. Based on what you know and what you can guess, answer the questions below. When you finish, you will share your predictions with other members of your team.

1. Where do you think you will live?
2. Will you live in a city or a rural area?
3. Will you be married?
4. Will you have children? How many?
5. How will you care for your children?
6. Will your parents live with you?
7. What sorts of food will you eat? Where will the food come from? Will you grow any of your own food?
8. What sort of work will you do? Where will you do it?
9. What sort of exercise will you get? Where will you go for it?
10. How healthy do you expect to be?
11. What will your surroundings be like? Crowded and polluted, or spacious and clean?
12. What other predictions can you make?

2. When students have finished, have them discuss and compile their responses in teams on chart paper, and post the results around the room.
3. After students have had a chance to read other groups' responses, discuss their predictions as a group. Questions to consider might include the following.
    - Are there any recurring themes or trends?
    - Are the outlooks generally positive or negative?
    - What do they think are some of the influences on their perceptions of the future?
    - Are there aspects of their predictions that they can personally influence? Which, if any, are out of their control?

Optional Extensions

1. Compare your class's responses to the questionnaire with students in other communities. What similarities and differences are there?
2. Draw a team mural depicting how you think your community will look when you are your parents' ages.

ACTIVITY 9.3

Water, Water Everywhere[3]

(Use with p. 105 in Chapter 6)

Contributed by: Ludmilla Payula
Physics, Maths Lyceum
School 239
St. Petersburg, RUSSIA

This activity introduces the students to the importance of water in their lives and some of the effects of water pollution. The activity's structure is a form of cooperative learning developed by a group of teachers in St. Petersburg, Russia. The activities are designed to be completed by teams of four students.

Objective

- Explore some of the fundamental properties of water

Materials

(per team) one copy of Instructions for Block 1; one copy of Instructions for Block 2; one set of cards A–D, for Blocks 1 and 2; 3 pairs of safety glasses; samples of petroleum products

If optional extension is to be pursued, one set of cards A–D for Block 3 and the following: copper sulfate crystals, vegetable oil, sugar, celsius lab thermometer, three capillary tubes of different diameters

Procedures

This activity consists of two "blocks." Each "block" consists of four different cards (one per team member). Each card contains three different pieces of information. Completion of a "block" consists of (1) individual work with one card, and (2) participation in a team activity involving all four cards.

[3] This activity is used with the permission of Ludmilla Payula.

Organization of Content Blocks

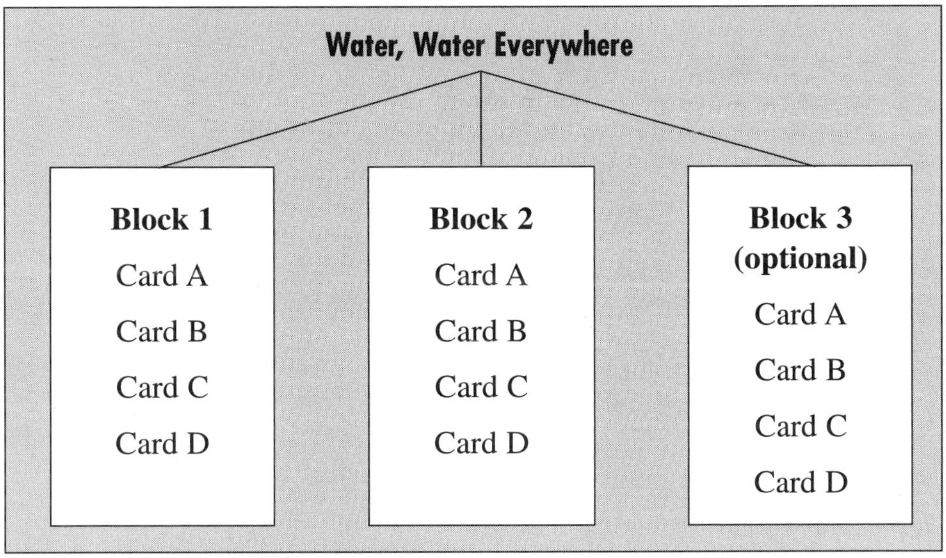

1. Distribute *Instructions for Block 1* from page 178 and one set of cards A–D from page 179 for Block 1 to each team. Allow time for team members to study the cards and share the information with their teammates.
2. Invite representatives from each team to report on their team's discussions to the class.
3. Distribute *Instructions for Block 2* from page 180 and one set of cards A–D from page 181 for Block 2. Allow time for team members to study the cards, and share the information with their teammates.
4. Invite representatives from each team to report on their team's discussions.

Optional Extension

A third set of activity cards (Block 3, A–D) address some of the special chemical properties of water. Have students follow the same procedures as Block 1. There are no follow-up questions.

Instructions for Block 1

Individual work with the cards:

1. Read #1 on the card. Think about how to explain its contents to your teammates. Use a reference book, if necessary.
2. Perform the task required in #2 on your card. Use a reference book to look up any unfamiliar terms.
3. Read the #3 on your card. It contains some practical advice or questions about the ecological situation in your city or your state.

Teamwork with the cards:

1. Each team member in turn should follow these steps:
   a. Read aloud #1 on your card to all members of your group. Use illustrations if necessary to explain the facts in full.
   b. Discuss your answer to question #2. If experiments were required, show them to your teammates, and discuss the results.
   c. Read aloud and explain #3 on your card to your teammates, and give an example.
2. After each team member has reported the information and activities on a card, discuss the following questions within your team. Prepare one team member to present this subject to the class.
   - What is the role of water in the lives of human beings?
   - What are the important life functions of water?
   - How much water is needed by a human being in 24 hours?

Block 1 (Cards A–D)

Block 1– Card A

1. Human beings consist primarily of water. Various body tissues contain different proportions of water. The total water content in the body of a young person amounts to 60–70 percent of the total body weight. Water makes cells elastic. All biochemical processes occurring in the body occur in aqueous solution. Body fluids also cushion organs such as the brain and help maintain organs in their proper positions in space. Fasting, a human being could lose all of its fat and almost 50 percent of its protein and still survive; a loss of 10 percent of its water is lethal.
2. What is the water weight in *your* body? How much water can you lose before endangering your life?
3. Water lost due to exertion or illness must be replenished. Remember that water loss also involves loss of dissolved salts. Therefore, it is important to replace lost body fluids with salt solutions rather than pure water.

Block 1 – Card B

1. The daily water requirement of an adult is 40 grams per kilogram of body weight.
2. Calculate your daily water requirement. In what ways is water normally lost by the body? Try to explain why a human being cannot live without water for more than a few days.
3. Dietitians recommend that we drink at least eight glasses of water each day.

Block 1 – Card C

1. Evaporating water facilitates body cooling. Since the heat of vaporization of water is high, evaporating water molecules remove a large amount of thermal energy. Many living organisms exploit this feature to cool their bodies.
   Many reactions that occur in the body are exothermic (they give off heat). Why are no "hot points" formed in the body? Why do you sweat when it is hot outside?
2. Why does your skin feel cool when alcohol is rubbed on the skin? If you were trying to reduce a fever, would it be more effective to use compresses of cold water or alcohol? Why?

Block 1 – Card D

1. One of the hypotheses explaining the origin of life suggests that life originated in the environment of the primordial ocean. Blood serum contains the same chemical elements, compounds, and ions as those found in sea water:

ions, compounds	blood serum	sea water
Na+	39.0	30.0
Mg_2+	0.5	3.8
Ca	1.0	1.8
K+	2.6	1.8
Ce-	45.0	55.2
CO_2	11.0	0.5
other elements	0.9	7.0

2. Why don't blood cells break if you drink a lot of water?
   Why don't blood cells shrink if you eat a lot of salty food?
3. Sodium and potassium ions are extremely important for human life. They are important for the normal functioning of nerve and muscle tissue. People with muscle cramps are sometimes encouraged to eat bananas, which are high in potassium.
   Nutrient substances such as mineral salts enter your body with food, and are absorbed into the blood and lymph through the large intestine. Blood carries nutrient substances to the body cells. The stability of the internal environment is maintained by the blood and internal organs.

Instructions for Block 2

1. Read #1 on your card. Think about how to explain it to a teammate.
2. Perform the task required in #2 on your card.
3. Read #3 on your card. It contains some practical advice or questions about ecological situations in your city or in your state.
4. Find one teammate. Teach your teammate the information in the first part of your card. Answer any questions.
5. Listen to the first part of your partner's card and ask questions.
6. Exchange cards and perform the task for #2 on the other person's card. Discuss the two tasks and your results with your partner.
7. Teach each other the information in #3 on your respective cards.
8. Find another member of your team to partner with and proceed as before. Your work is finished when you have completed steps 4–7 with each member of your team.
9. After your team has completed its work with the cards, discuss the following questions. Prepare one team member to present the results of your discussions to the class.
   - What kinds of water pollution have you learned about?
   - Can you think of other examples of how humans pollute water?
   - How can polluted bodies of water affect the surrounding areas?
   - How can bodies of water be protected from pollution, while allowing for technological development?

Questions for Block 2 (Cards A–D)

Block 2–Card A

1. Phosphorus is an element that encourages plant growth. Phosphorus-containing fertilizers greatly increase crop yield. Lakes and rivers get their phosphorus from rocks, and their phosphorus levels are usually low. Lakes are polluted by phosphorus by means of waste water runoff, which is rich in phosphorus-containing fertilizers and detergents.
2. What do you think would be the effect of phosphorus pollution on a lake? Could trout live in such a lake? (Trout prefer clean water rich in oxygen.)
3. The processes occurring in lakes polluted by phosphorus cannot be stopped, but they can be slowed. If the waste flow is stopped, algae feeding on the pollutants will die and sink to the lake bottom. Slowly, the lake will clear. What do you know about the lakes in the vicinity of your town? Do trout live in them? Do they contain surplus phosphorus?

Block 2–Card B

1. Acid rain is a very serious problem. Acid rains are formed when sulfur-containing compounds (mainly sulfur dioxide) discharged into the atmosphere by industrial plants and automobiles react with water vapor in the air to form acids. These acids precipitate to the Earth's surface as rain and snow. Acid rain destroys paint and stone, as well as damages the environment of lakes and ponds. Various bodies of water can become so acidic that fish die.
2. Write a chemical equation for the formation of acid rain.
3. A 1979 survey carried out in the Adirondack Mountains of New York State indicated that all fish had died in 264 lakes in the area due to acid rain. What sort of effects do you think this had on the local economy?

Block 2–Card C

1. Many toxic substances are known to be dangerous for aquatic organisms. Most of them are absorbed and accumulated by these organisms. Fish living in a lake poisoned by chlorohydrocarbons and mercury will store these chemicals in their bodies, thus becoming poisonous for people to eat. Plastic and aluminum debris also pose dangers for aquatic animals and humans.
2. Given time, it is possible that marine microorganisms can decompose plastic. Since plastic is similar in composition to nutrients, what would be the result of this process? Would this water be drinkable?
3. Mariners report that the Atlantic surface waters from the Caribbean Sea to the coasts of Europe are littered with plastic debris. What would be your policy with regard to plastic and aluminum? How would you describe the bodies of water near where you live? Are they polluted with plastic and metals?

Block 2–Card D

1. Petroleum products may greatly pollute the water environment. Oil is less dense than water; it is not water soluble. It has less surface tension than water, which is why it spreads on the surface, forming oil slicks that cannot be penetrated by oxygen. Petroleum products pollute coastal areas and beaches, endanger the existence of living organisms, and kill birds living along the coast. Oil spills are often the result of tanker wrecks, oil exploration in the offshore zone, and ship diesel pollution.
2. Petroleum products are refined to get kerosene, heavy oils, gasoline, lubricants, and other products. You should check their behavior in water. It is similar to that of petroleum. Fill three glasses with water, and then add a few drops of kerosene, gasoline, or heavy oil to each. What changes do you see in the different glasses?
3. Near where you live, are the bodies of water polluted by petroleum products? Can you identify any of the sources of this pollution?

Questions for Block 3—Optional Extension

Block 3–Card A

1. The unique properties of water are determined by chemical structure. The oxygen atom is covalently bound to two hydrogen atoms. The molecule is bent, with an oxygen atom at its apex, and two hydrogen atoms at the sides. Since oxygen attracts electrons more strongly, the oxygen atom has a partial negative charge, while the hydrogen atoms have partial positive charges. The oxygen atom of one molecule is attracted to the hydrogen atom of another, forming a hydrogen bond. Each water molecule may become hydrogen bonded to four other water molecules.
2. For a molecule of its small size, water has a relatively high boiling point. Can you explain why, taking into account its molecular structure?
3. Most of the solar energy absorbed by the Earth's surface is lost due to evaporation. Solar energy absorbed by the vaporization of water is released again when vapor condenses into clouds. This energy becomes thermal energy and warms the air. Atmospheric moisture can be compared with a warm blanket covering the Earth.

Block 3–Card B

1. The water molecule's polarity is responsible for its most interesting property: its ability to dissolve various compounds. The most soluble substances are ionic compounds, e.g., salts and other polar compounds. The process of a solution dissolving is accompanied by heat absorption and release. Many aqueous ions are colored. Non-polar substances dissolve poorly in water. In these cases, an interface is formed between water and the non-polar compound. These interfaces are very important in living organisms, since it is on such interfaces that many chemical reactions occur.
2. Fill three small glasses with water and measure the water temperature. Add some copper sulfate crystals to the first glass, and stir carefully. Note whether the compound dissolves and whether there is a temperature change. Repeat with vegetable oil and sugar. In which glass did you produce aqueous ions? Why do you suppose it is recommended that salad oil be poured on a salad immediately before serving?
3. Mineral water is healthful. It contains a lot of dissolved salts and other compounds vital to living organisms. Are there mineral spas or springs in your state? Where? Have you ever visited one or tasted the water?

Block 3–Card C

1. Water molecules are subject to cohesive forces, which attract them to one another. The water molecules at the surface of water are attracted down into the water. This results in a sort of "skin" on the water's surface called a surface tension film. (Try floating a needle on the surface of a glass of water, and you will be able to see this "skin.") Water also adheres to other materials and "wets" them. "Wetting" and surface tension explain the ability of water to move into small pores and cracks, even when opposed by gravity.
2. Place three capillary tubes of different diameters vertically in a glass of water. In which tube is the water level highest? Why? Why does the volume of your hair seem to decrease when it is wet?
3. Various additives to water can decrease its surface tension. Float a needle on top of a glass of water. Carefully add a few drops of detergent near the edge. What happens?

Block 3–Card D

1. Water is a chemically active substance. It participates in millions of chemical processes occurring in the natural environment. For example, rainwater containing dissolved carbon dioxide reacts with limestone (calcium carbonate) to produce soluble calcium bicarbonate. When this solution comes to the surface and evaporates, calcium carbonate precipitates again, forming stalactites.
2. If calcium bicarbonate and magnesium are present in water, the water is said to be "hard." What are some disadvantages of having "hard" water in your home? Find out how water "softeners" work.
3. Stalactites are among the most beautiful of tourist attractions. Are there caves in your area that contain interesting limestone formations? Have you ever visited them?

CHAPTER 10

The Student Log

The GTP Student Log should be used to encourage your students to write about global thinking and to record data and information about environmental topics that they study. Each activity in a GTP project has corresponding pages in this log.

Here are some ways to use the GTP Student Log:

- Photocopy the entire log for every student, and as you do the activities in the GTP projects, have students write about their environmental science ideas.
- Photocopy the pages of the log that correspond to the project or the individual activity your students are working on.
- Provide one copy of the log pages to each group rather than each individual student.

Add some blank pages to the end of the Student Log packets for field notes and journal entries. We hope the student log enhances your students' experiences with the Global Thinking Project.

The Global Thinking Project Student Log

Welcome to the Global Thinking Project

Dear Students,

We want to welcome you to the Global Thinking Project, a telecommunications project linking students together by means of computers and satellites for the purpose of getting to know each other better while they study environmental science. There are students from many countries around the world participating in the project, including students from Australia, the Czech Republic, New Zealand, Russia, Singapore, Spain, and the United States.

During the year you will work not only with students in your own class, but with students from around the world. You will use the Internet to stay in contact with your Global Community by sending electronic mail (e-mail) reports of your work to each other. You will also use the GTP Web site to learn about the environment, share information that you collect yourself, and find resources to help you discover how best to solve environmental problems.

What topics will you study? The GTP is organized around a series of environmental science projects. Projects Hello and Green Classroom will begin your study, followed by Project Clean Air. Later in the year your class will study one or more projects, including Project Solid Waste, Project Water Watch, and Project Soil. In April, the entire Global Thinking Project will work on Project Earth Month. For Project Earth Month, your class will design and carry out an environmental science project that has special meaning to your city or town. You will develop a Project Earth Month Web page and have it linked to the Global Thinking Project Web site.

In the pages that follow, you will find information about the GTP projects, which you will use to write about global thinking and to record data and information about the environmental topics you study.

Best wishes for a successful and fruitful year. And remember to *Act Locally, and Think Globally.*

Jack Hassard, Ph.D.
and Julie Weisberg, Ph.D.
Coauthors

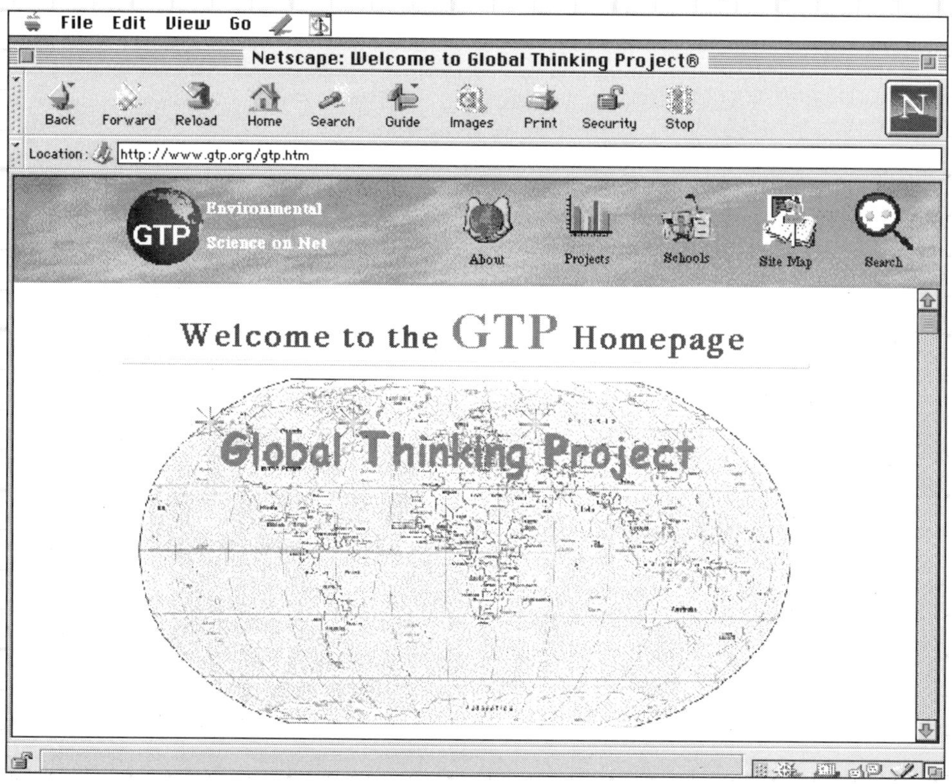

The Global Thinking Project Home Page: www.gtp.org

What Is Global Thinking?

Locate Global Thinking Schools on the world map.

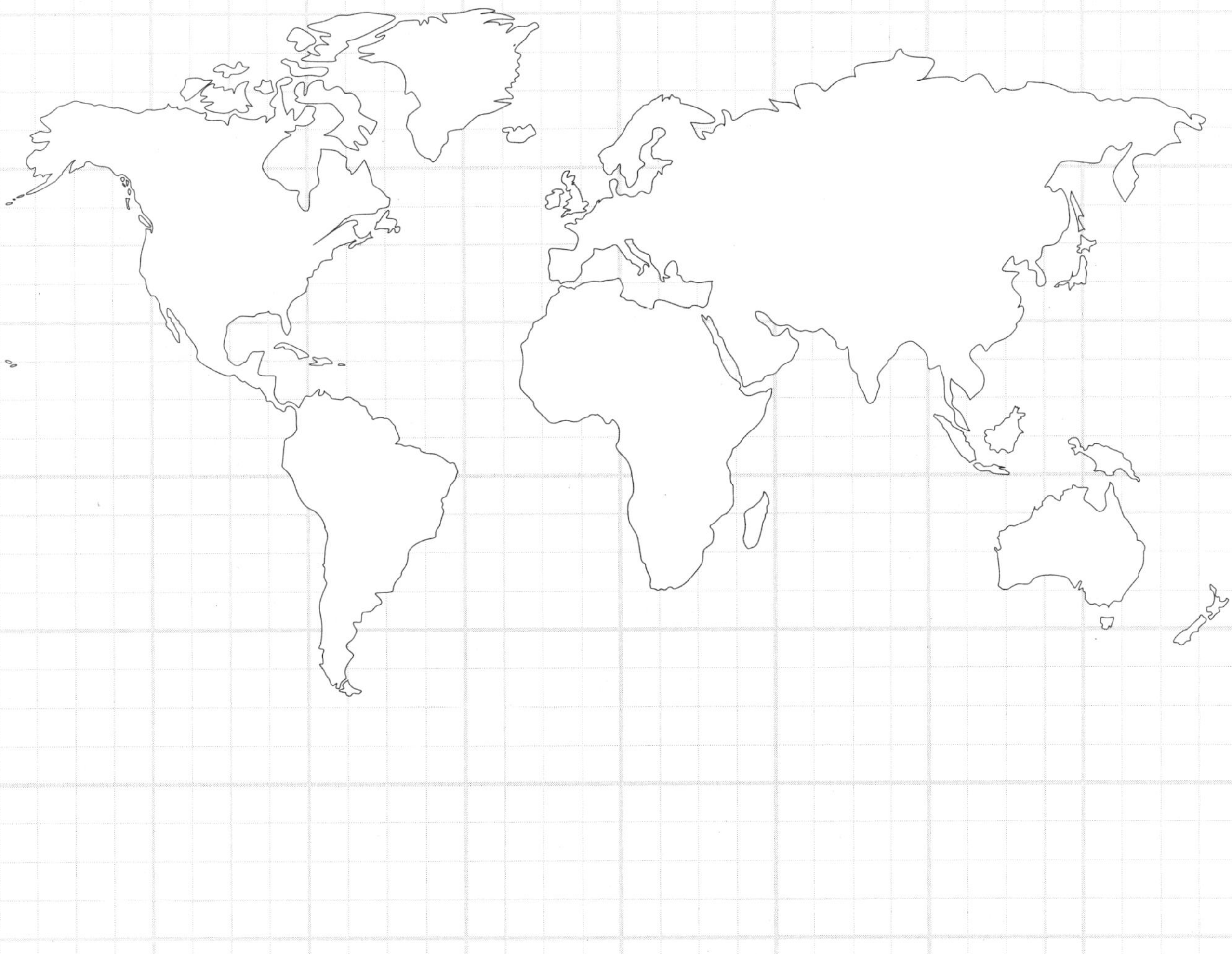

The definition of “global thinking,” in my opinion, is . . . .

A Hello Message

Draft a "hello" message.

Class Hello Pacs

What did your class include in the Hello pacs? Why did you decide to include these things?

Earth as Living System

To which resource group were you assigned? ______________________

How does this resource contribute to the maintenance of the other resources, and of plant and animal life?

Sketch your team's model of the interrelationships in the biosphere.

Global Environmental Problems

Write your team rankings of global problems, followed by the class's rankings.

	Our Team	Class Average
war technology		
hazardous substances		
extinction		
water resources		
human health		
population growth		
air quality		
mineral resources		
world hunger		
land use		
energy		
nuclear reactors		

Explain your team's rankings.

How did your team's rankings compare with the rest of the class?

How Will Global Problems Change by 2020?

For each global environmental problem below, students will indicate whether they think the problem will be better, about the same, or worse, by the year 2020.

Environmental Problem	Much Better or Better	About the Same	Worse or Much Worse	Don't Know
War Technology (nerve gas, high-tech weaponry, nuclear arms threat)				
Hazardous Substances (waste dumps, toxic chemicals, lead paint)				
Extinction of Plants and Animals (reduction of genetic diversity)				
Water Resources (waste disposal, estuaries, supply, distribution, ground water contamination, fertilizer contamination)				
Human Health (infectious and noninfectious disease, stress, diet and nutrition, exercise, mental health)				
Population Growth (world population, immigration, carrying capacity)				
Air Quality and Atmosphere (acid rain, carbon dioxide, depletion of ozone layer, global warming)				
Mineral Resources (non-fuel minerals, metallic and non-metallic minerals, mining, technology, low-grade deposits, recycling, reuse)				
World Hunger and Food Resources (food production, agriculture, cropland conservation)				
Land Use (soil erosion, reclamation, urban development, wildlife habitat loss, deforestation, desertification)				
Energy Shortages (synthetic fuels, solar power, fossil fuels, conservation, oil production)				
Nuclear Reactors (nuclear waste management, breeded reactors, cost of construction, safety)				

Project Green Classroom

Weather Team Assignment

You will need a Celsius lab, thermometer, sling psychrometer, and a camera with a light meter.

Your task is to make some simple measurements that will allow you to make some statements about the classroom's climate over the course of a day.

1. Measure the air temperature of the room. Do you think it is the same everywhere in the room? How could you find out? Does distance from the floor matter? Devise a way of measuring the air temperature at a constant height at different points in the room. Is the air temperature in the classroom constant during the day? Measure the air temperature at the same place in the room at intervals during the day.
2. Use the sling psychrometer to measure the relative humidity in the classroom. If you can, make these measurements throughout the day, at the same intervals at which you measure the air temperature.
3. Use the light meter to measure the light intensity in the room. If the classroom has many windows, you might want to make several measurements at different distances from them.
4. Measure the air temperature, relative humidity, and light intensity just outside your classroom. How similar or different are the indoor and outdoor readings?

Weather Team Data Table

	Outside	**Location #1 Inside**	**Location #2 Inside**	**Location #3 Inside**
Time #1 Temperature Rel. Hum. Light Int.				
Time #1 Temperature Rel. Hum. Light Int.				
Time #1 Temperature Rel. Hum. Light Int.				

Air Quality Team Assignment

You will need particulate collectors (prepared in advance by your teacher), microscope or hand lens, mm ruled graph paper.

Your task is to examine particulates that have been collected from the air in your classroom over a period of a week.

1. Examine the slides that have been placed around the room during the past week to collect particulates. Using a microscope or hand magnifier, determine
   - the average number of particles per square mm
   - the types of particles (small, medium, large; range of colors; range of shapes)
2. Are there differences in the results depending on where the collectors were located? What hypotheses can you make to account for differences? If you wish, put out several new collectors to test your hypotheses.

Air Quality Team Data Table

	Location 1	Location 2	Location 3
# or Particles/mm^2			
Particle size . . . #1g/mm^2 # med/mm^2 #small/mm^2			
Particle shape(s)			
Particle color(s)			

Physical Plant Team Assignment

You will need paper and pencils, measuring tapes.

Your task is to map the physical parameters of the classroom.

Measure the length, width, and height of the classroom. Calculate its volume. Draw a scale floor plan of the classroom. Indicate locations of windows, doors, heating/air conditioning vents, furniture, plants, and anything else visible in the room. How many square feet of wall are devoted to windows? (What fraction of the total wall space is this?)

What type(s) of light fixtures are used? How many hours a day are they used? How many kilowatt hours of electricity are consumed lighting your classroom? How much does this cost the school system?

Physical Plant Team Data Table

Classroom Measurements		
Length of room (meters)	________	m
Height of room (meters)	________	m
Width of room (meters)	________	m
Volume of room (meters^3)	________	m^3
Total wall area (meters^2)	________	m^2
Area of windows (meters^2)	________	m^2
Percent of wall space that is windows	________	%
Lighting		
Type of fixtures		
# of each	________	
# of hours used in one day	________	
# kwh electricity consumed	________	

Sketch of the room:

Population Team Assignment

You do not need any special equipment.

Your task is to determine how the classroom is used by macroscopic living things.

1. How many human beings occupy this room during a day? How long do they stay? count the number of people who go in and out of the classroom during the day, and how long each stayed. Calculate the number of person-minutes the classroom was occupied.
2. How many plants are there in your classroom? Of what types? Are they large or small? Do they appear to be healthy? (What criteria are you using to assess their health?)
3. How many animals are there in your classroom? Of what types? Are they large or small? Do they appear to be healthy? (What criteria are you using to assess their health?)

Population Team Data Tables

Human Population

Time Interval	# of People	Person-minutes
Total time:	Total # of people:	Person-minutes:

Plant Population

Type	Number	Size	Health

Animal Population

Type	Number	Size	Health

Microorganism Team Assignment

You will need agar plates exposed in different parts of your classroom (prepared in advance by your teacher), dissecting microscope or hand lens, reference books.

Your task is to examine the types of microorganisms that can be collected in different parts of your classroom.

Without opening them, examine the microorganisms growing in the Petri dishes. Sketch and describe the different types of colonies you observe. Use reference books to determine whether your colonies are mold or bacteria (mold colonies are generally fuzzier and larger than bacterial colonies). Were different microorganisms present in different areas of the classroom? Develop some hypotheses for your results.

Microorganism Team Data Table

	Location 1	**Location 2**	**Location 3**	**Location 4**
Sketch of colonies				
Description of colonies				
Identification of colonies				

Trash Team Assignment

You will need one day's trash collected from your classroom, paper sacks, digital bathroom scale.

Your task is to investigate the trash produced in your classroom during a typical day.

1. Sort the trash collected for one entire day in your classroom into categories: paper and paperboard, metals, glass, plastics, rubber and leather, textiles, food wastes, other.
2. Place each collection in a paper sack and weigh it on the bathroom scale. (One way to do this is to weigh one team member with and without each sack.
3. Calculate the total mass of trash collected in your classroom in one day, and determine the percent of each category of trash present.

Trash Team Data Table

Category of Trash	Mass (grams)	% of Total Mass
Paper		
Metal		
Glass		
Plastic		
Rubber/leather		
Textiles		
Food wastes		
Other		
TOTAL		100

Green Classroom Data Chart

Category	Information	Your Class	School 1	School 2	School 3	School 4
School	name					
Town	name					
Date	day-month-year					
Latitude	hh: mm n/s					
Longitude	hh: mm e/w					
Temp. Inside	Average of temperatures					
Temp. Outside	Average of temperatures					
Relative Humidity	%					
Light Intensity	meters					
Room length	meters					
Room height	meters					
Room width	meters					
Room volume	cubic meters					
Population	Number of people in room/day					
Plants	Total number in room					
Animals	Total number					
Particles	Average number of particles/mm					
Particle size	Average particle size in room					
Micro-organisms	Yes, if some observed; No, if not					
Paper	% of total					
Metal	% of total					
Glass	% of total					
Plastic	% of total					
Rubber/ leather	% of total					
Textiles	% of total					
Food	% of total					
Other	% of total					
Trash	Total Mass in grams					

Briefly describe the results of the following teams:

Weather Team:

Air Quality Team:

Physical Plant Team:

Population Team:

Microorganisms Team:

Trash Team:

What did you learn about the environment of your classroom?

How can you and your classmates improve the environment of your classroom?

How does your classroom environment compare to schools in other communities?

A Breath of Clean Air

What we know about air pollution:

What our class has heard about air pollution	What questions our class has about air pollution

Answer the following questions.

1. How does the quality of the air impact you directly?

2. How does it affect other living things in the environment?

3. What is "clean air," and what causes the air's quality to change?

4. How do you think the air quality in your region compares to that in regions close to you? in places far from you?

Monitoring the Air

Keep a record of your class weather data in the chart below.

Row	Air Temp.	Rel. Hum	Cloud Cover	Sun-shine	Wind Speed	Wind Direc.	Precip.	Lat.	Long.	School	Date
1											
2											
3											
4											
5											

Contents of Your E-mail Letter:

Monitoring Particles

Do a Web search and find answers to the following questions. Include the Web address.

Why are we concerned about particle matter?
Web address:
Answer:

Who is most at risk from exposure to fine particles?
Web address:
Answer:

What is particulate matter and what are "fine" particles?
Web address:
Answer:

How do particulate matter and fine particles affect the environment?
Web address:
Answer:

What have you heard about particulates?

What are some examples of particulates?

How can you detect or observe particulates?

Describe the experiment your class did to measure particulates.

Monitoring Ozone

What have you heard about ozone?

Keep a record of your class's ozone data in the chart below. Record data from other schools as you receive e-mail.

Observations	Day 1	Day 2	School	School	School	School
Quantitative Data						
Ozone						
Temperature						
Wind Speed						
Wind Direction						
Precipitation						
Latitude						
Longitude						
Qualitative Data						
Ozone						
Temperature						
Humidity						
Clouds						
Sunshine						
Wind Speed						
Wind Direction						
Precipitation						
Particulate Count						

Project Clean Air Recording Form

Team Name: ____________________ Team Members: ____________________

Measurement Location: Inside ____________________ Outside ____________________

Date: ____________________ Latitude: ____________________

Start Time: ____________________ Longitude: ____________________

Quantitative Measurements

Ozone (Eco Badge® Readings)

Location	Start Time	End Time	Reading 1 ppb	Reading 2 ppb	Reading 3 ppb	Average ppb

Temperature: ____________________ °C

Relative Humidity: ____________________ %

Wind Speed: ____________________ km/hr

Wind Direction: ____________________ degrees

Precipitation: ____________________ cm

Qualitative Observations

Circle one of the words in each row to describe your observations.

Ozone	Good	Moderate	Unhealthful	1st Stage Alert
Temperature	Hot	Warm	Cool	Cold
Humidity	High	Average	Low	
Clouds	Cloudy	Partly cloudy	No clouds	
Sunshine	Sunny	Mainly sunny	Mainly cloudy	Cloudy
Wind Speed	Strong	Medium	Light	Zero
Wind Direction	North	East	West	South
Precipitation	Rain	Drizzle	Snow/sleet	None
Particulates	High	Moderate	Low	

Other Observations: ____________________

____

Web Discussion:

What did you learn about air pollution when you read other GTP students' comments on the Project Clean Air home page?

Air Pollution: What Can You Do About It?

Choose from the following action projects to make your own contribution toward cleaner air.

- Ozone Web site
- Newsletter
- Video Program
- Poster Report
- Research Paper
- Other—Identify Your Project

Write a brief statement describing the focus of your study.

Getting to Know Your Trash

Household Waste Survey

1. For each category, record the weight of the garbage you collected. Add to get the total daily garbage output of your family.
2. Compute the percent of the total weight contributed by each category.

Category	**Mass of Garbage**	**% of Weight**
Paper and paperboard		
Metals		
Glass		
Plastics		
Rubber and leather		
Textiles		
Yard wastes		
Food wastes		
Other		
Totals		

How does your trash at home compare to your trash in school?

How does trash data from your class compare to those from other schools in your global community?

The Three Rs: Reduce, Reuse, Recycle

Read the article below and participate in a debate about solid waste.

The Garbage Glut Is Everyone's Business

Source: *USA Today*, November 17, 1989[1]

Garbage just won't go away. And we can't ignore it anymore. There are fewer and fewer places todump it.

The "pooh pooh choo choo"—5,000 tons of sewage sludge—is chugging its way back to Baltimore after a month-long trip, first to Donaldsonville, La., where residents turned up their noses, then to Moss Point, Miss., where the welcome mat was also yanked.

Five train cars of rotting refuse from Philadelphia headed for Brazil, Ind., met the same fate last month. A court order turned it around and sent it back home. And in Los Angeles, trash trucks hustle from one dump to another, trying to find one that hasn't taken in its daily quota. Unless something is done by 1992, the city will be turning out 6,000 tons of garbage a day with no place to put it.

Across the USA, city and county officials are longing for the good old days, when the mountains of garbage produced by conspicuous consumption were whisked away, never to be whiffed again. But garbage that used to be dumped, burned, then buried is now burying us. Landfills, where most trash winds up, are disappearing at an alarming rate. The Environmental Protection Agency expects that 2,000 of them will disappear within four years. What's worse, landfills have not proved to be the busy little biodegraders that people expected. They've actually preserved hot dogs and carrots for 15 years; 20-year-old newspapers can still be read.

What can be done? Plenty. We can produce less trash. More firms are finding ways to reduce packaging waste. And the federal government has set a goal of reducing garbage production by 25% by 1992. We can convert more trash to energy. Tazewell County, Ill., hopes to turn methane gas at its landfill into electricity for 2,400 homes. More communities need to seek ways to convert waste to power.

And we can recycle more of our trash. Thirty states are implementing laws to separate trash for recycling; 10 have mandatory separation. The recycling idea is catching on. Two-thirds in a recent poll said they had recycled some garbage. Minnesota dairy farmers are finding that shredded newspapers make great cow bedding. School cafeterias and fast-food restaurants earn money by recycling plastic foam products. New Yorkers return three-quarters of their bottles and cans for deposits each year, keeping 100 million tons of trash out of landfills.

Such efforts show that individuals can make a huge difference. But the writer across the page thinks the price is too high. He'd rather pay someone else to do his sorting. The price may be steeper than he thinks. In Woodbury, N.J, he'd pay a fine of up to $500 a day for not sorting his trash. In Hamburg, N.Y., he would forfeit trash pickup. The price is steep because the consequences are dire: We can be buried in our own refuse if we don't act soon. Ignoring garbage won't make it go away. Dealing with it will.

It's a dirty job, but everybody has to do it.

[1] "The garbage glut is everyone's business," *USA Today*, November 17, 1989. Copyright 1989, *USA Today*. Reprinted with permission.

How Long Will It Be There?

Make a list of the items your team decided to bury to investigate decomposition rates.

Rank in order the items that will decompose at the fastest rate to the one that will decompose at the slowest rate. Give reasons for your choices.

Evaluating Waste Disposal Options

Explain how an environmental problem can become an environmental issue.

List values on which you think environmental issues decisions should be based.

Evaluate the effectiveness of the following waste disposal options by evaluating them according to criteria based on your values. Record your criteria and ideas in the chart below.

Criteria				
Alternatives:				
Open Landfill				
Sanitary Landfill				
Incineration				
Source reduction				
Recycling				
Other				

Explain your reasoning for your selection.

Where Does Your River Come From; Where Does It Go?

Make a sketch showing the path your drinking water takes from the local water supply to its tributaries.

Dirty Water Clean Up Challenge!

After your teacher shows you the “dirty water,” work with your cooperative group to come up with a method to clean up the water. Record your work below.

Write down the questions your team decides to answer. Your question might begin with “What will happen if we . . . ?”

Write down how you are using the materials to conduct experiments.

Use this space to record any data you collect. You might want to make data tables.

Summarize your results. What method do you recommend to clean up the dirty water?

Send and e-mail message to other students in the GTP by reporting your results on the Project Water Watch home page.

Field Trip No. 1: Observing the Physical and Chemical Characteristics of the River

What do the following test results tell you about the nature of a river?

Level of dissolved oxygen

ph level

Rate of water flow

Depth of river

Temperature of river

Describe the river you visited. Where is it located? How big is it? Where does it come from? Where does it go?

Water Watch Observation Form

Team Name: ______________________ Team Members: ______________________

Measurement Site (circle one): river, creek, pond, lake, ocean, other: ______________________

Date: ______________________ Latitude: ______________________

Weather: ______________________ Longitude: ______________________

Quantitative Measurements

Air Temperature:

Water Temp.: ________ #1 ________ #2 ________ Avg. ________

Depth: ________ #1 ________ #2 ________ #3 ________ Avg. ________

Flow: ________ #1 ________ #2 ________ Avg. ________

D. Oxygen: ________ #1 ________ #2 ________ Avg. ________

pH: ________ #1 ________ #2 ________ Avg. ________

Qualitative Measurements (Circle one word in each row.)

Land use	wooded	meadow	park	residential	commercial
Erosion	none	gullies			
Ground cover	grasses	shrubs	trees	soil	
Odor	rotten egg	musky	oily	sewage	none
Algae color	n/a	light green	dark green	brown	blue
Bed-bottom color	gray	orange/red	yellow	black	brown
Surface water	clear	oily	foamy	milky	muddy
Discharge	yes	no			

Notes: ______________________

____

____

____

Water Watch Observations

Sketch a segment of the river (or body of water). Include as many details as possible.

Draw a cross section of the body of water. Include as many details as possible, including the location of plants and rocks.

Making Sense of the River Data

To make sense of the data you collected and observation you made at the river, it is helpful to organize the information in a data table.

Use each row for a set of data from not only your school, but from the other schools in you global community.

Quantitative Data

	Temp.	Time	Date	Site	School	Lat.	Long.	Depth	Flow	D. O_2	pH
1.											
2.											
3.											
4.											

Qualitative Information

	Land Use	Erosion	Ground Cover	Odor	Algae Color	Bed-Bottom Color	Surface Water	Discharge
1.								
2.								
3.								
4.								

Activity

Using the information from the Quantitative Data, make a data table using the Alice Network software, or a graphics software program. Analyze the data by making graphs and summaries of the information.

Field Trip No. 2: Collecting and Identifying Aquatic Macroinvertebrates

Use the form below to record observations of macroinvertebrates that you and your teammates collect at the river. Use the identification chart on the next page to classify your finds.

Aquatic Macroinvertebrates Monitoring Survey

Group: ____________________ Group Members: ____________________

Location: river: ________ creek: ________ pond: ________ lake: ________ ocean: ________

Date: ____________________ Weather: ____________________

Air Temperature: ____________________ Water Temperature: ____________________

Depth of Water in Collection Site: ____________________

Sketch of Collection Site:

Group One Taxa: Pollution-Sensitive Organisms; signify good-quality water		**Group Two Taxa: Somewhat Pollution-Tolerant Organisms**		**Group Three Taxa: Pollution-Tolerant Organisms; can be in poor-quality water**	
Organism	**Count**	**Organism**	**Count**	**Organism**	**Count**
Stonefly nymph		Crayfish		Aquatic worms	
Caddisfly larva		Sowbug	larva	Midge	
Water penny beetle larvae		Scud		Blackfly larva	
Riffle beetle				Leech	
Mayfly nymph				Pouch snail	
Gilled snail		Danselfly nymph		Other snails	
Dobsonfly larva		Crane fly larva Dragonfly nymph Clam			
Total Group 1		Total Group 2		Total Group 3	

Using Macroinvertebrates as Bioindicators

Analysis of a River's Health

Using the chart on the next page, determine the stream quality based on the macroinvertebrates that you collected and identified.

1. Excellent
2. Good
3. Fair
4. Poor

What were some of the organisms that you found? List them and draw what they looked like.

Name	Drawing

Prepare an e-mail letter to your global community. Use the e-mail template below when writing your e-mail letter.

To: Global Community Schools
Re: Macroinvertebrate Report

Here are the results of our river study of macroinvertebrates.

<date>, <location>, <weather>, <air temperature>, <water temperature>, <depth>, <number of group one taxa organisms>, <number of group two taxa>, <number of group three taxa>, <latitude>, <longitude>, <school name>, <town>

We also found that: ______________________________

____

____

____

____

____

____

____

____

Signed

Stream Quality Assessment: Observations and Analysis

The organisms in Figure 1 are grouped into three categories:

GROUP 1	(pollution-intolerant or good quality indicators)
GROUP 2	(organisms that can exist in both extremes of quality)
GROUP 3	(pollution-tolerant or poor quality indicators)

The organisms in these three groups are assigned a group index value.

GROUP 1 = 3 GROUP 2 = 2 GROUP 3 = 1

The analysis procedure consists of counting the number of types of organisms in each category and multiplying the group index value.

EXAMPLES:	**GROUP 1 TAXA**	**GROUP 2 TAXA**	**GROUP 3 TAXA**
	CADDISFLY(S)	DRAGONFLY(S)	BLACKFLY(S)
	STONEFLY(S)	CRAYFISH	MIDGE (S)
	MAYFLY(S) 3x3=9	CLAM(S) 3x2=6	2x1=2

The respective group index values are then added together to find the cumulative index value (which in the above case would be 17). By referring to the following chart, the stream quality assessment can thus be determined.

STREAM QUALITY ASSESSMENT	**CUMULATIVE INDEX VALUE**
EXCELLENT	23 and above
GOOD	17 - 22
FAIR	11 - 16
POOR	10 or less (SEE ACTION PROCEDURE ON PAGE 8)

NOTE: The organisms on the stream quality assessment form used in the field are to be recorded by placing a letter code in the corresponding block. Each letter represents your estimated count.

Protecting Our Rivers

Reflect on what you have learned about rivers.

How can rivers be protected?

Are there rivers in your community that need to be protected? Why?

What Is Soil?

What is soil?	What questions do you have about soil?

Below are five soil forming factors.

1. Parent Material: the type of rock from which soil is formed affects its chemical composition and texture
2. Climate: important climatic influences on soil formation are precipitation and temperature
3. Organisms: plants and animals affect soil formation as decomposers and consumers of organic matter, and as earth movers
4. Topography: the degree of slope of a site may influence the quantity of precipitation absorbed, the rate of soil removal by erosion, and the movement of suspended and dissolved materials from one area to another
5. Time: weathering of rock and the subsequent modification of soils by organic processes takes place over relatively long periods of time

Field Examinations of Soil

Team Name ______________________________ Date ______________________________

Brief description of the site:

Sketch of the site:

Sketch of the pit, with horizons shown:

Data Table

	Horizon #1 (from surface)	**#2**	**#3**	**#4**
top boundary (cm from surface)				
bottom boundary (cm from surface)				
color				
structure				
consistence				
texture				

Laboratory Investigations of Soil

Soil Moisture

Sample #	No. 1	No. 2	No. 3	No. 4
Mass of beaker				
Mass of beaker + wet soil				
Mass of wet soil				
Mass of beaker + dry soil				
Mass of water in soil sample				
% of water in the original sample				

Soil Texture

Trial #	No. 1	No. 2	No. 3
starting volume of soil			
volume of sand settling after 40 seconds			
volume of sand and silt settling after 30 minutes			
volume of silt			
volume of clay in original sample			
% sand			
% loam			
% clay			
type of soil (from textural triangle)			

(Repeat, if desired, with samples from different soil layers.)

Soil Acidity

Sample #	No. 1	No. 2	No. 3	No. 4
pH (trial 1)				
pH (trial 2)				
pH (trial 3)				

Textural Triangle[2]

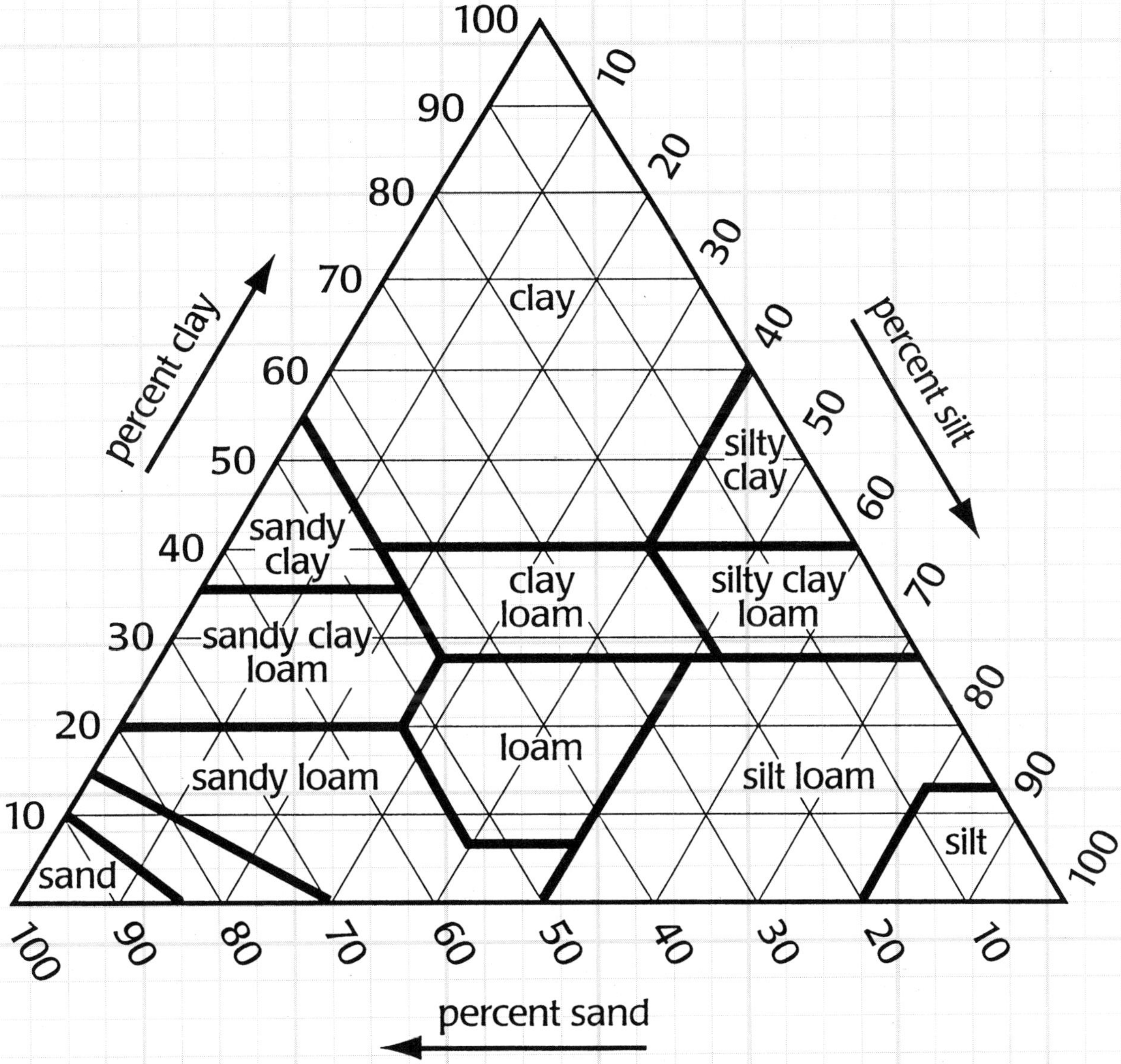

2 Reprinted by permission of The G.L.O.B.E. Program, www.globe.gov, 1-800-858-9947.

Preserving the Soil

My notes from our tour around the school grounds:

Some ideas for school-improvement projects:

Our Action Plan:

Project Earth Month

Project Earth Month is an action-taking activity of the Global Thinking Project. During the year, students have been learning how to investigate environmental problems, and how to use telecommunications to collaborate with students around the world. During Earth Month, which traditionally is held during the month of April, the Global Thinking Project invites your class to become involved in environmental action strategies including the following.

Persuasion: convincing others, individuals and groups, that a suggested course of action is correct

Political Action: bringing pressure on political or governmental agencies and/or individuals in an effort to influence positive environmental action

Eco-management: taking action to maintain a good physical environment or to improve a weakened physical environment

Our Earth Month Project

Your class will be participating in a global event. Each school in the GTP during the month of April will identify and work on an environmental problem or issue, and then design a project to help resolve the questions asked. To help you complete the task, refer to the chart on the next page to help guide your work.

Steps for Action-Taking Projects

Stage	Description	Teacher's Role	Student's Role
1. Determine subtopics and organize into groups	Students scan sources, propose questions for an environmental project.	Introduce general topic of investigation; lead groups in discussion of topic and subtopics.	Generate questions of interest; sort questions into categories; form project groups.
2. Plan group projects	Group members plan project together; decide what they will study, and how they will go about it. They divide the work up among themselves.	Help groups formulate plans; meet with groups; help find resources and provide equipment and materials.	Plan what to study; choose resources; assign roles; send e-mail report describing plans.
3. Carry out projects	Group members organize and analyze information and data, pool their findings, and form conclusions. Members discuss their work in progress in order to integrate ideas.	Help with study and research skills; check with groups to maintain progress.	Seek answers to question; locate information; collect data; interview people; integrate findings and summarize.
4. Plan presentations	Members determine the main idea of their presentation and how to present their findings.	Discuss with class plans for presentations; organize the process.	Determine the main idea for the presentation; plan how to make presentation; prepare materials for presentation (poster, video, etc.).
5. Make presentations	Presentations are made to class and others.	Coordinate presentations; insure that e-mail reports are sent to Global Community and gtp.earthconf.	As presenters, give feedback to classmates; send e-mail reports; read and react to e-mail reports from other schools.
6. Evaluate	Students share feedback about the projects. Projects are evaluated by teachers and students.	Evaluate learning and summarize the projects.	Reflect on work as a group member; reflect on impact of project on environmental literacy.

What are some environmental issues or problems that you think the class should study? Make a list of several in the space below.

What topic is your class going to study? What do you think about this topic? What do you know about the topic?

Keep track of the work your team and class does on Project Earth Month. Keep a record of your work in your journal.

Resources

Books

Brooks, Paul. *Speaking for Nature: How Literary Naturalists, from Henry Thoreau to Rachel Carson, Have Shaped America.* Boston: Houghton Mifflin, 1980.

Chiras, Daniel D. *Environmental Science: A Framework for Decision Making.* Menlo Park, CA: Benjamin-Cummings, 1985.

Corson, Walter H., ed. *The Global Ecology Handbook: What You Can Do About the Environmental Crisis,* by the Global Tomorrow Coalition. Boston: Beacon Press, 1990.

Cousteau, Jacques and Cousteau Staff. *The Cousteau Almanac: An Inventory of Life on Our Water Planet.* New York: Doubleday, 1981.

Cummins, K.W. and M.A. Wlizbach. *Field Procedures for Analysis of Functional Feeding Groups of Stream Macroinvertibrates.* Appalachian Environmental Laboratory Contribution No. 1611, University of Maryland, 1985.

Earth Works. *50 Simple Things Kids Can Do to Save the Earth.* Berkeley, CA: Earth Works Press, 1991.

The Global Tomorrow Coalition. *The Global Ecology Handbook.* Boston, MA: Beacon Press, 1990.

Lehmkuhl, Dennis. *Aquatic Insects.* Burr Ridge, IL: WCB McGraw Hill, 1979.

Hollender, Jeffrey. *How to Make the World a Better Place.* New York: William Morrow, 1990.

Lewis, Barbara A. *The Kid's Guide to Social Action.* Minneapolis: Free Spirit Press, 1991.

Miller, G. Tyler. *Living in the Environment: An Introduction to Environmental Science.* Belmont, CA: Wadsworth Publishing Company, 1990.

Rifkin, Jeremy, ed. *Green Lifestyle Handbook.* New York: Henry Holt, 1990.

State of the World (published annually). Worldwatch Institute, 1776 Massachusetts Avenue, NW, Washington, DC 20036

Wilke, Richard J. *Environmental Education Teacher Resource Handbook: A Practical Guide for K–12 Environmental Education.* Milwood, NY: Kraus International Publications, 1993.

Organizations

Population Growth

Zero Population Growth
1400 16th Street, NW
Washington, DC 20036
(202) 332-2200

Food and Agriculture

World Resources Institute
1709 New York Avenue, NW
Washington, DC 20006
(202) 638-6300

Biological Diversity and Extinction

National Audubon Society
950 Third Avenue
New York, NY 10022
(212) 832-3200

National Wildlife Federation
1400 16th Street, NW
Washington, DC 20036-2266
(202) 797-6800

World Wildlife Fund
1250 24th Street, NW
Washington, DC 20037
(202) 293-4800

Young Entomologist Society, Inc.
International Headquarters
1915 Peggy Place
Lansing, MI 48910

Tropical Forests

The Nature Conservancy
1815 North Lynn Street
Arlington, VA 22209
(703) 841-5300

Ocean and Coastal Resources

Marine Debris Information Office
Center for Marine Conservation
1725 DeSales Street, NW
Washington, DC 20036
(202) 429-5609

The Cousteau Society
930 West 21st Street
Norfolk, VA 23517
(804) 627-1144

Fresh Water

American Rivers
801 Pennsylvania Avenue, SE, Suite 400
Washington, DC 20003
(202) 547-6900

Save Our Streams
Izaak Walton League of America
1401 Wilson Boulevard, Level B
Arlington, VA 22209
(703) 528-1818

Natural Resources/Energy

National Arbor Day Foundation
211 North 12th Street, Suite 501
Lincoln, NE 68508
(402) 474-5655

National Energy Foundation
5160 Wiley Post Way, Suite 200
Salt Lake City, UT 84116
(801) 539-1406

Union of Concerned Scientists
26 Church Street
Cambridge, MA 02238
(617) 547-5552

Air/Atmosphere/Climate

Acid Rain Foundation, Inc.
1410 Varsity Drive
Raleigh, NC 27606
(919) 828-9443
Provides resources on acid rain, air pollutants, and global climate change.

Environmental Defense Fund
257 Park Avenue South
New York, NY 10010
(212) 505-2100

Hazardous Substances

National Toxics Campaign
29 Temple Place, 5th Floor
Boston, MA 02111
(617) 482-1477

Solid-Waste Management/Recycling

Keep America Beautiful, Inc.
9 West Broad Street
Stamford, CT 06902
(203) 323-8987

Environmental Action

Kids Against Pollution
P. O. Box 775 High Street
Closter, NJ 07624
(201) 768-1332

Kids for Saving the Earth
620 Mendelssohn, Suite 550
Golden Valley, MN 55427
(612) 525-0002

Take Pride in America
P. O. Box 1330-Y
Jessup, MD 20794

Earth Day

Earth Day USA
P. O. Box 470
2 Elm Street
Peterborough, NH 03458
(603) 924-7720
For a small fee, Earth Day USA provides a comprehensive Earth Day Planning Packet containing ideas, resources, and suggestions for planning an Earth Day event.

Internet and Software Providers

Mindspring

(888) 677-7464

NETCOM

San Jose, California
(408) 983-5950

Institute for Global Communications (IGC)

1731 Connecticut Avenue, NW
Washington, DC 20009-1146
(202) 588-5070

The Testbed for Telecollaboration (TERC)

2067 Massachusetts Avenue
Cambridge, MA 02140
(617) 547-0430
testbed@teaparty.terc.edu
Alice software developer

Project Materials Suppliers

Carolina Biological Supply Co.

2700 York Road
Burlington, NC 27215
(800) 334-5551
Sells nutrient agar plates

Chemetrics, Inc.

Calverton, VA 22016
(540) 788-9026

Forestry Suppliers, Inc.

P.O. Box 8397
Jackson, MS 39284-3770
(800) 647-5368
Sells light meters, Dwyer wind meters, standard soil color charts

Kraus Publications

(800) 223-8323
Publisher of the *Environmental Education Teacher Resource Handbook*

The Science Source

P.O. Box 727
Waldoboro, ME 04572
(800) 299-5649
(207) 832-7281 (fax)
www.thesciencesource.com
Sells the student psychrometer, item #16220

United States Department of Agriculture
14th Street and Independence Avenue, SW
Washington, DC 20250
(202) 720-2791
Soil surveys

Vistanomics, Inc.
230 North Maryland Avenue
Suite 310
Glendale, CA 91206
(818) 409-9157
(818) 409-9334 (fax)

Internet Sources

Environmental Protection Agency Web site
www.epa.gov

Global Thinking Project Web site
www.gtp.org

Russian GTP Web site
www.education.spb.ru/gtp/

GTP site at the A. Herzen State Pedagogical Institute, St. Petersburg, Russia
www.emissia.spb.su/gtp/gtp.html

Glossary

Air pollution: any toxic or radioactive gases and particulate matter introduced into the atmosphere, principally as a result of human activity. Pollutants can be carried by winds and rising air currents, with the larger particles returning to the Earth's surface under gravity, while smaller particles are removed by rain.

Alice: a software package that incorporates data gathering, graphing, and mapping. The Alice software can be used on Macintosh or IBM computers. It was developed by The Testbed for Telecollaboration (TERC).

Chlorofluorocarbons (CFCs): a family of gases used as propellants in aerosol spray cans and as refrigerants. CFCs have been found by scientists to drift into the upper atmosphere and destroy ozone.

Cloud: visible masses of water or ice that has condensed on tiny particles in the air. Visit the Clean Air Web page in the GTP Web site to learn more about clouds, their formation, and types.

Conference: sometimes called a bulletin board, a conference is a "meeting place" on the network in which users can discuss particular topics. Each project home page has its own bulletin board.

Eco Badge®: a chemical system used to measure the amount of ground-level ozone in the air.

E-mail: electronic mail; e-mail is similar to an ordinary letter except that it is transmitted via computer.

Environmental issue: an environmental problem about which there is a disagreement about the solution.

Environmental problem: a problem that is the result of the interactions between humanity and the environment.

GAIA: a virtual character designed to interact with Global Thinking Project participants.

Ground level ozone: ozone gas that forms near the ground due to the combination of hydrocarbons and nitrogen oxides in periods of high temperature and sunlight.

Humidity: the amount of moisture in the air.

Hydrocarbons: a chemical compound that contains carbon and hydrogen atoms, such as natural gas.

Incineration: disposal of wastes by burning.

Internet: the name of a group of worldwide information resources; a huge source of practical information available by computer and modem.

IP (Internet Protocol) Address: an address assigned to your computer when connected to the Internet. It usually has four numbers separated by periods, e.g., 144.188.311.6.

Modem: short for modulator/demodulator; a device that converts (modulates) the electronic signals that your computer and the network understand into signals that can travel over telephone lines, and then converts (demodulates) them back again.

Nitrogen oxides: a chemical compound that contains nitrogen and oxygen atoms.

Off-line: work done when your computer is not connected to the network.

On-line: work done when your computer is connected to the network.

Open landfill: a large area of land that is excavated and filled with waste.

Ozone: a highly reactive gas comprised of three oxygen atoms (O3). Ozone is produced by the recombination of oxygen in the presence of ultraviolet radiation from the sun. Ground level ozone is an atmospheric pollutant that is formed by the reaction of hydrocarbons and nitrogen oxides in the presence of high temperature and sunlight.

Ozone hole: a space or hole in the ozone layer caused by destruction of ozone gas by chemicals drifting to the upper atmosphere.

Ozone layer: the thin layer of ozone gas found in the upper atmosphere that shields living things from ultraviolet light.

Ozone shield: a layer of ozone in the upper atmosphere that protects life on Earth by filtering out harmful ultraviolet radiation from the sun.

Particulate matter: solid particles and liquid droplets present in the atmosphere. Particulate matter has various sources, both natural and as a result of human activity. Examples include soot and ash emitted from factories, dust raised by agricultural and mining activities, and dust and ash generated by volcanic eruptions.

Photochemical reactions: these are reactions among gases in the air, initiated by the energy from sunlight.

Photons: a tiny amount of light energy.

Precipitation: moisture from the air in the form of rain, hail, sleet, snow, frost, dew.

Primary pollutant: any substance released into the atmosphere in harmful concentrations.

Secondary pollutant: a pollutant resulting from the chemical reaction between two or more primary pollutants or a primary pollutant and a naturally occurring atmospheric element.

Recycle: the extraction of useful materials from waste for reuse. Examples of recyclable materials are aluminum, batteries, corrugated cardboard, clothing, glass, paper and cardboard, plastics, oil, tin and steel cans, tires, and yard waste.

Reduce: a movement to reduce solid wastes by decreasing the number and volume of discarded products. This includes reducing the amount of unnecessary packaging and reducing the amount of toxic chemicals in these products.

Reuse: a movement to "precycle," meaning consumers purchase long-lasting, reusable items, rather than disposables.

Sanitary landfill: a landfill that is lined with a nonporous material to prevent liquids from leaching into the surrounding soil, and in which each day's deposits of waste are covered with a layer of dirt.

Smog: a dense, discolored fog containing large quantities of soot, ash, and gaseous pollutants.

Soil: a mixture of weathered rock and decaying organic matter that covers the surface of the Earth and serves as a growth medium for plants.

Soil Horizons: distinct layers of soil with different properties that are parallel to the Earth's surface. A formal classification for soil horizons, which distinguishes primarily organic from primarily inorganic layers, may be found in soil science texts.

Soil Properties: soil characteristics that distinguish one soil from another. These may include color, texture, consistence, and structure.

Telecomputing: the process of transmitting messages or signals over a distance by means of computers connected to telephones via modems.

Temperature: a measure of the degree or intensity of heat or cold.

Ultraviolet light: very short wavelengths of light that are harmful to living things. Most UV light is filtered out by ozone in the upper atmosphere.

Web-assisted teaching: using the resources of the Internet as a part of your lesson plans.

Wind: the horizontal movement of air from higher air pressure to lower air pressure.

About the Authors

Jack Hassard, Ph.D.

Jack Hassard is a professor in the Department of Middle and Secondary Education and Instructional Technology at Georgia State University in Atlanta, Georgia. He has directed the activities of the Global Thinking Project from its inception and continues in that capacity. He has authored many books and curricula in science and science education and presents seminars and workshops at national and international levels. He can be reached by e-mail at jhassard@gsu.edu.

Julie Weisberg, Ph.D.

Julie Weisberg is associate professor in the Department of Education at Agnes Scott College in Decatur, Georgia. She is the co-director of the Global Thinking Project, and has worked on the project from the beginning. She is the author of papers and articles on science education and global thinking and is active in professional meetings and conferences at national and international levels. She can be reached by e-mail at jweisberg@agnesscott.edu.

Acknowledgments

Many educators, students and parents have contributed to the creation of the Global Thinking Project (GTP). It is impossible to name all of them, but they represent citizens from many parts of the world. Each has contributed in their own way to foster the ideas and aspirations of the GTP.

We are especially appreciative of more than three hundred American and Russian families that participated in the GTP–Georgia/Russia Student/Educator Exchanges from 1995–1998. These American and Russian families participated with their schools in yearlong exchange programs that focused on global thinking and environmental science research.

We also wish to thank the funding agencies which supported this project, including the Eisenhower Higher Education Program in Georgia, the U.S. Environmental Protection Agency, the United States Information Agency, and the Continuing Education Division of Georgia State University. The project was supported also by Georgia State University, Agnes Scott College, the Russian Academy of Education, as well as schools and school districts in the United States and abroad.

We wish to express our appreciation to the following educators and colleagues for their contributions: **Australia:** Julie Brown, Roger Cross, Jenny Carter Cross, Paul McColl, Barry Plant, Kelli Sewell, Julian Cross, **Czech Republic:** Roman Hasek, Zdenek Novak, Vaclav Nydl, Pavel Stroner, Jiri Vanicek, **Russia:** Raisa Alexeyeva, Elena Antonenkova, Vladimir Baskunov, Natalia Bogdanova, Kseniya Balobanova, Ludmila Bolshakova, Tatyana Bukina, Nadezhada Feruloyova, Galina Ghuravskaya, Marina Goryunova, Anatoly Karpov, Tatyana Khokholva, Yuvanali Koulutkin, Marina Kurbatova, Galina Manke, Svetlana Mansurova, Elena Martynova, Alexei Matushkin, Alexander Martynenka, Nadya Nevpokoeva, Nataliya Novozhilova, Olga Olenynikova, Alexander Orlov, Ludmila Payula, Tatyana Puskaeva, Dmitry Razumny, Tatyana Ruban, Vera Ryzhova, Lyubova Samarina, Galina Soukhobskaya, Galina Temchukova, Sergei Tolstikov, Simeon Vershlovsky, Tatyiana Vasilyeva, Irina Yakimanskaya, Inna Zabrodina, Anatoly Zakhlebny, Valentina Zalim, Galina Zhuravskaya, Varvara Zommerova; **Singapore:** Tang Woh Un; **Spain:** Joan Adam, Miguel Aguareles, Ramon Barlam, Joan Berrocal, Rosa Casas, Albert Cervera, Joan Castella, Carme Espinar, Carme Lliviuna, Anna Pinyero, Julia Martin, Ricad Martin, Rosa Ochoa, Monste Rodon, Angels Serentill, Joanna Soler, Josep Torrents, Narcis Vives; **United States:** Robert Almeder, Michael Anderson, Deborah Ault-Butenko, Marna Barnard, Boris Berenfeld, Stewart Brewster, Barbara Broadway, Ron Broadway, Harvey Brodsky, Leslie Carroll, Ted Colton, Pascale Creek-Pinner, Sarah Crim, Margaret Coomer, Sherita Daniels, Helen Davis, John Davis, Sam Dietz, Sylvia Dodson, Susan Dunkerly-Kolb, Augustin Dunn, Marcia Eidelman, Margaret Fredrico, Phil Gang, Cheryl Garner, Wanda Gordon, Tom Greening, Edith Guyton, Mary-Alice Hassard, Alan Hoffman, Marcia Humbert, Diane Jacobi, Gary Lieber, Jill Kelly, Joel King, Harris Lintini, Fran Macy, Bob Maxfield, Dorothea McAlvin, Martha McIlveene, Trinna McKay, Martha Mersereau, Dori Miyashita, William Nye, JoAnn Polascik, Gail Pappacostas, Pat Perfetti, Robert Pike, Mary Rigger, Wayne Robinson, Peggy Sherry, Gary Short, Dave

Smiley, Lee Sisson, Beck Spotts, Rose Talbert, Roy Tarr, Katherine Taylor, Margaret Venable, Nelly Van Os, Nancy Walton, Catherine Warfield, Derek Whordley, Jane Yonts.

We would especially like to acknowledge the contributions of Yong Hu, a graduate student at GSU, whose skillful programming brought the GTP Web site from concept to reality. Finally, we wish to thank Roberta Dempsey for encouraging us to publish this book, and Laura Strom and Susan Buntrock for an incredible job preparing the manuscript for publication.

Unless otherwise acknowledged, all photographs are the property of Addison-Wesley Educational Publishers, Inc. Page 21 NASA; p. 43 Mark Richards/Photo Edit; p. 55 courtesy of the Suquamish Museum; p. 103 USDI; p. 127 Ron Sherman; p. 147 David Young-Wolff/Photo Edit.